SIDEWAYS UNCORKED

PHOTO BY MERIE WEISMILLER WALLACE/© 2003 FOX SEARCHLIGHT /PHOTOFEST

SIDEWAYS UNCORKED

The PERFECT PAIRING of FILM & WINE

Kirk Honeycutt and Mira Advani Honeycutt

APPLAUSE
THEATRE & CINEMA BOOKS

Dedicated to:
Cinephiles, Pinotphiles & Merlot Mavericks

Bloomsbury Publishing Group, Inc.
ApplauseBooks.com

Distributed by NATIONAL BOOK NETWORK

Library of Congress Cataloging-in-Publication Data Available

ISBN 978-1-4930-7804-2 (pbk.: alk. paper)
ISBN 978-1-4930-7805-9 (electronic)

♾️™ The paper used in this publication meets the minimum requirements of American National Standard for Information Sciences—Permanence of Paper for Printed Library Materials, ANSI/NISO Z39.48-1992

Contents

Introduction

Sideways is now two decades old. Somehow it doesn't feel that long ago. Then again, it feels like ages. It feels like a different era when an indie film, seen in a motion picture house and not in a living room or on a "device," could bring so many diverse folks together to laugh at such unlikely characters getting involved in something many moviegoers knew very little about and no American feature had ever explored before—the joyous world of wine.

The film's director, Alexander Payne, had a substantial following thanks to three comedies before *Sideways*, so a modest success looked likely. The movie was about the kind of things that movie executives normally distrust, featured characters deemed marginal, starred mostly unfamiliar actors, and showcased a region and a wine little known to most viewers. So no one—not the filmmakers nor the actors nor anyone in the film or wine businesses—was prepared for what happened.

The box-office performance of the comedy, bringing in $110 million worldwide after topping out at $71.5 million domestically, doesn't begin to tell the story, as staggering as those numbers are for such a modestly budgeted independent picture with limited aspirations. The film sharply impacted the lives of many involved with the film; recalibrated the American wine industry; made a star of a red wine few knew existed, namely Pinot Noir; became part of the zeitgeist of that era; and sent shock waves through various wine regions, especially the comedy's main location, California's Santa Barbara County. And, infamously, it damaged—for a while—the fortunes of a highly popular red wine called Merlot.

The movie world glories in its unlikely hits, in underdog movies that outperform expectations. But few if any of these have impacted another business to such a degree. Nothing prepared the American wine business for the tsunami that followed. Only a handful of vintners and restaurateurs in the Santa Barbara area were even aware of the movie before its release. The wine industry never got the memo.

The comic tale of two rascally, roving buddies meandering in California wine country made wine hip. It gave wine personality. It looked like fun. And it demonstrated how well wine, especially Pinot Noir, pairs with romance.

How did this happen? How did Paul Giamatti and Pinot Noir become household names?

The genesis for this book came about on July 31, 2022, when the authors attended a crowded open house at what is now known as Clendenen Lindquist Vintners (CLV), a thirty-thousand-square-foot wine facility in the foothills of the famous Bien Nacido Vineyards in Santa Maria, California. This funky, cavernous, no-frills warehouse was started back in 1989 when maverick vintner Jim Clendenen and his longtime friend and fellow vintner Bob Lindquist moved into a much smaller facility to create a home for several local winemakers.

We hadn't visited the winery in several years to enjoy one of Jim's daily family-style lunches cooked by him in an open kitchen, owing to COVID and busy schedules. For all the fabulous Santa Barbara wines to taste and a huge buffet to eat that afternoon, it was a melancholy occasion for us. Jim passed away suddenly in his sleep over a year before at age sixty-eight. All the CLV winemakers share this facility to make and house their wines, but Jim was the heart, soul, and, it should be added, "Mind Behind" the joint. On every label of Jim's wine, Au Bon Climat (ABC), you will see in small print the name of Jim Clendenen followed by "Mind Behind." That was his handle.

Other pioneer winemakers came to Santa Barbara before Jim, but Jim was one of those guys, like Robert Mondavi in Napa Valley, who brought recognition to a wine region. Jim and Au Bon Climat elevated the wine style and culture of the community. A tireless promoter, he brought Santa Barbara Pinots and Chardonnays to the world, not just to North America. A larger-than-life personality with shaggy blond hair and loud shirts covering his considerable girth, Wild Man Jim, as he was known, preached the sermon of balance between alcohol and acidity to produce a wine that will enhance a meal.

We visited the tasting table of Frank Ostini, winemaker and restaurateur at The Hitching Post, whose wines and eatery were so celebrated in the movie that he says the movie gave him "immortality," not to mention a fortune in sales. We struggled to find seats amid the throngs of oenophiles. As we thought about the array of Pinot Noirs Frank and others now make thanks to the movie, we reflected on that movie's incredible reach. It was unlikely that anyone in that gathering had not seen the movie at least once.

In 2004 we attended the world premiere of *Sideways* at the Toronto International Film Festival. Kirk Honeycutt, then the chief film critic for *The Hollywood Reporter*, had already raved about *Sideways* in his online review that day.

The next January we sat at the table of Alexander Payne and his cowriter Jim Taylor at the Los Angeles Film Critics Association awards banquet, where Kirk handed those two gentlemen the award from the organization for best screenplay of 2004. Later, Mira Honeycutt convinced Chronicle Books to let her write a guide to California's Central Coast wineries because of the growing *Sideways* phenomenon. Jim Clendenen wrote the book's foreword.

A term coined many years ago was "the *Sideways* effect." It was deployed primarily by wine journalists to denote the effect that the movie had on the business they covered, but it extends to those involved with the movie, the Santa Barbara wine region, and the changed manner in which many people now think about wine and wine culture.

How and why did this happen? What was it about this oddball comedy with its *Mutt and Jeff* characters, one in a fight against the existing world, his own self, and the mediocrity if not mendacity of certain wines, and the other caught up in a foolish slipstream of irresponsible behavior in response to a failure to grow up?

We thought we knew some of the answers, but we knew we didn't know them all. We realized this would be a book about two things—movies and wine. Two seemingly unconnected subjects until *Sideways* connected them. You can no longer talk about the movie without considering the wine. Together they make a fascinating tale of how popular culture works, how movies pervade our lives and imaginations, and how mavericks in both fields, cinema and wine, sent us in new directions with their relentless search for new stories. Yes, movies and wines tell stories. They are fascinating and rich with personalities searching for their voices.

This book should be enjoyed with a glass of Pinot Noir. Or two. Toward that end, we have designated a few selections from various regions with a nod toward availability and price. There's little point in recommending a rare bottle worth four figures. That's not what real wine tasting, the kind Miles and Jack indulge in, is all about. Wine is for everyone.

So this book is for Jim and all the Santa Barbara wine pioneers, for Alexander and his team, and for all the ambitious, relentless, risk-taking vintners who refused to listen to naysayers when it came to Pinot Noir and the filmmakers who refused to compromise their vision of the perfect wine movie.

Cheers!

1

Ode to Indie Filmmaking

The movie *Sideways* is an independent film. Yet if you look up the 2004 film online, you'll see that *Sideways* was fully financed by Fox Searchlight, then a wholly owned subsidiary of Twentieth Century Fox. Fox was then a major Hollywood studio. So how can this movie be considered "independent"?

One thinks of an independent film as a low-budget movie shot and financed with private money that a producer then takes to the marketplace—say, the Sundance Film Festival or a film market—to sell to a distributor. Such a process does happen, but it's much less common. As we'll see, *Sideways* was presold to Fox Searchlight before principal photography began.

Unlike most art forms, movies are *all* about money. Money is their mother's milk. As tough as it is to get movies made within the Hollywood studio system, meaning where production, distribution, and marketing costs are borne by the major motion picture distributors, it's that much tougher to make and distribute genuinely original, nongeneric, nongroupthink works. These films, almost always lower in budget and almost always higher in inspiration and imagination, have long been labeled "independent."

So what is meant by the term "independent film"? Is it another kind of branding? A marketing ploy? Or does it describe a process that is crucial for guarding a filmmaker's autonomy? Who, after all, is independent? All moviemakers are dependent on money.

If you ask Alexander Payne, the director and cowriter of *Sideways*, if he's an indie filmmaker, he sighs.

"I don't know, man," he says. "I consider myself a filmmaker. I guess 'independent' refers to the nature of films I've made. I never call myself an independent filmmaker. We have to define 'independent.'"

Yes, we should define what we mean by that term, because it's the very nature of the films Alexander Payne makes that is the crux of the matter here. It's simple: he makes the films he wants to make. If he wants to shoot *Nebraska* in black-and-white, something the studio looked on with bafflement verging on horror, he takes Directors Guild of America scale, a mere $225,000 for three years of work, to bend the suits to his will. If he wants to cast Paul Giamatti and Thomas Haden Church as the two guys footloose in Santa Barbara wine country in *Sideways*, instead of movie stars such as George Clooney and Brad Pitt, he not only does so but he still gets the budget he needs to shoot on location out of town.

So let no one doubt there is a thing called independent cinema. Let's hold onto that designation and see if we bring about some clarity.

The landscape of American cinema has always included independent filmmakers. Film pioneer Samuel Goldwyn set himself up as an "independent producer" back in 1923, and his production-only operation (he had no distribution arm) remained that way for thirty-five years. His company functioned as a studio (with contract players and employees) and earned its share of Oscars, but he still insisted on the independent label.

When MGM's top-earning director King Vidor decided to go freelance and wanted to make a film during the Depression about a farm cooperative where the unemployed could work out a subsistence living, every studio and then every bank passed. These capitalist gentlemen refused to back a film that edged too close to the Soviet ideal of collective farms for their comfort.

So Vidor hocked his house, automobile, and anything else of value to make *Our Daily Bread*—independently. His good friend Charlie Chaplin agreed to release the picture in 1934 through United Artists, a company he co-owned. So Vidor, too, one of the biggest directors of that era, became an "independent." Only no one used that term.

Small production companies have come and gone across the history of American cinema. A cluster of indie movie studios on or near Gower Street in Hollywood earned the slang term "Poverty Row" for their extremely low-budget operations grinding out Westerns and other genre pictures, sometimes referred to as "B pictures," from the presound era into the 1950s. These certainly weren't major studio pictures but, again, no one bandied about the term "independent."

Yet as early as the mid-1950s one of the first film historians, Arthur Knight, noticed that directors increasingly had an alternative to the studio system, that being what he called "independent production," with each film individually made, financed, and marketed, which he noted in his seminal 1957 film history, *The Liveliest Art*.

The New Wave films of the 1950s and 1960s in the United States included avant-garde and so-called underground cinema. You could not expect to find them at your local movie house but rather on college campuses or at cinema societies. No one knew what to call them. "Underground" stuck for a while, as there was something so ephemeral about them, things that briefly surfaced, blinked at the unaccustomed light, then vanished overnight.

These countercultural filmmakers such as Kenneth Anger, Andy Warhol, Maya Deren, Jonas Mekas, and Stan Brakhage were "independent" from traditional media on several fronts: technological (often using 8 mm and 16 mm films, unlike the industry's ubiquitous 35 mm), aesthetic (personal, avant-garde, or experimental as opposed to conventional), financial (artistic without little or any monetary motive), and sociopolitical (focusing on marginal people in American culture).

Then came the movie that changed everything.

In 1958, a Hollywood actor, John Cassavetes, an iconoclastic Greek American who starred in movies such as *Edge of the City*, *The Dirty Dozen*, and *Rosemary's Baby*, released a movie called *Shadows* made on the streets of New York for little money (reportedly $15,000, but exact figures are hard to come by). He shot the film in 16 mm in 1957, then reworked it in the editing room. It was scripted by Cassavetes and rehearsed by the actors, yet it felt improvisational.

Among other things, the film about a love affair between a white boy and a black girl tackled the subject of race, a virtually taboo subject then in Hollywood. This first feature launched what people started calling "the New American Cinema," in which actors need to inhabit their characters and work improvisationally (or at least *seem to*) in finding emotional truth.

After two frustrating forays into studio filmmaking, Cassavetes made *Faces* (1968), about a couple whose marriage is on the brink of collapse. Again, he shot in 16 mm and black-and-white (later blown up to 35 mm) and edited for about two years. He thus evolved a style that deeply influenced many filmmakers to come, including Martin Scorsese, Robert Altman, and Elaine May.

***

Gradually film journalists, critics, and filmmakers developed an ongoing conversation around the phenomenon everyone called the New American Cinema. Everyone knew there were these unusual films, often made in regions in America where studio films never ventured, dealing with subjects and issues Hollywood swore never to touch. Unfortunately, there was another defining characteristic—a cineast had a damn difficult time trying to *find* these films.

You might catch a title or two at a film festival, on college campuses, or, ironically, in Europe, especially on television. For years indie filmmakers talked about organizing an informal network of US distributors and exhibitors to get their pictures into movie houses where people could see them.

In 1979 the Independent Feature Project (IFP) was born to support regional filmmakers by providing access to funding and cultivating exposure to wide audiences. The impact of this organization on these emerging maverick American filmmakers was huge.

"Our whole function is to make sure a filmmaker doesn't have to reinvent the wheel every time someone decides to distribute his or her film," declared Joy Pereths, a serious-minded young Englishwoman with a background in distribution, who was appointed the new association's executive director.

The association had a long-range goal of centralization of distribution procedures but *not* production—"regional production is the key to this strategy," she affirmed, tellingly—the dissemination of information regarding financing, distribution, and exhibition so members can learn from each other's experiences, and finally lobbying for additional public and private funding.

She said this in 1980. In the years since, IFP has helped foster early works by such filmmakers as Michael Moore, Jim Jarmusch, Ed Burns, Kevin Smith, and Mira Nair, among many others.

Since its founding, the group has split into two to cover the East and West Coasts. IFP/West launched its Independent Spirit Awards in 1986—the indie Oscars, if you will, now a staple of the awards season—and the East Coast version, later rebranded the Gotham Film & Media Institute, launched its own Gotham Awards in 1991 to showcase and honor films made in the northeastern region of the United States. IFP/West later became Film Independent.

The increasing popularity of cable television and the advent of affordable home video players in the 1980s opened up new markets for these maverick filmmakers. Bobby Roth remembers finding himself in deep debt when he made his debut feature *The Boss' Son* in 1978. But he was able to sell the film to Los Angeles's famous Z Channel, one of the nation's first pay television stations.

"It wasn't only Z Channel," he recalls. "They paid little. OnTV and SelecTV paid more. The three pay channels got me whole. I paid off my investors."

Roger Corman, who began his career making snatched-from-the-headlines exploitation movies for American International Pictures, created in 1970 the first of his own production companies. He hired young filmmakers such as Martin Scorsese, Francis Ford Coppola, John Sayles, Peter Bogdanovich, James Cameron, Gale Anne Hurd, Ron Howard, Jonathan Demme, and Joe Dante, and actors such as Jack Nicholson, Bruce Dern, David Carradine, Peter Fonda, Dennis Hopper, and Robert De Niro to produce over four hundred films on shoestring budgets, and nearly every one saw a profit. These films featured rampaging bikers, vicious mobsters, vengeance-seeking strippers, blood-lusting vampires, and armies of aliens. Subtlety was not his strong suit.

A close second to Roger Corman was a couple of ambitious Israeli cousins, Menahem Golan and Yoram Globus, who purchased a financially ailing film company, Cannon Films, and went on to produce over three hundred films before running out of steam. Some, such as *Runaway Train* (1985) and *Street Smart* (1987), even got Oscar nominations.

These indies as much as film schools trained the new directors. As Alexander himself notes, "We who trained in the eighties were either in film school or worked for Corman or Cannon."

***

An audience began to develop for American-made independent films, as well as European "art" films, that were more dialogue driven and featured editing patterns that, while not necessarily challenging Hollywood continuity cutting, went at a more leisurely pace; they didn't propel viewers through a conventionally told story but rather lingered to observe characters' quirks and uncertainties.

Enough of an audience emerged for these nontraditional movies that the major studios began forming "classics" distribution divisions to target this budding and very loyal demographic, the first being United Artists Classics in 1980, which handled both foreign and American indie movies.

The biggest contributor to the contemporary independent film movement, though, is a thing called the Sundance Film Festival. Like the movement itself, the festival started small and virtually unnoticed. The first US Film Festival, the precursor to what we now know as the Sundance Film Festival, took place in Salt Lake City in September 1978. It touted itself as a festival of "regional cinema" where filmmakers were making money appealing to audiences the studios generally ignored.

Admittedly, many of these regional filmmakers were imitating Hollywood. George A. Romero, a commercial and industrial filmmaker in Pittsburgh, raised

$60,000 from friends in 1968 to shoot a black-and-white zombie horror film, *The Night of the Living Dead*, that became a cult favorite on the college and midnight circuits, grossing $12 million. At the inaugural US Film Festival, Romero brought a follow-up to his cult classic in a modern-day vampire story called *Marvin*.

The grand prize winner was Claudia Weill's *Girlfriends*, a regional film from the region of New York City about two young women living a life on the margins of Manhattan.

What happened to Weill was, unfortunately, the norm then. Despite fine critical notices, her film flopped commercially. It opened in New York the week the city's newspaper strike began. Then, in other cities, it ran into a problem many indies face—no "hook," such as a major star, and the enormous cost of advertising, which wipes out the box-office take in one or two theaters.

A retrospective program at that US Film Festival featured Joan Micklin Silver, like Alexander Payne an Omaha native, who made a series of wonderful indie films such as *Hester Street*, *Between the Lines*, *Chilly Scenes of Winter*, and *Crossing Delancey*. Working with her husband Raphael Silver, she made her first film, *Hester Street*, about a group of Russian Jewish immigrants in turn-of-the-century New York, in 1975 for $350,000. Unlike most indie filmmakers even today, Joan and Ray Silver self-distributed that film, with profitable results.

Then, in 1980, the well-known Hollywood film director Sydney Pollack (*The Way We Were*, *Out of Africa*) made a suggestion, or so the legend goes, to this still nascent festival. Why not move the US Film Festival up to the mountains, to Park City, in the wintertime? A festival taking place in a ski resort during ski season would bring Hollywood scouts, acquisition executives, and distributors by the droves, he argued. The festival's organizers took heed, and the event slotted into the second and third weeks of January, traditionally the slowest times of the ski season. It was relaunched as the United States Film and Video Festival. It would not officially become the Sundance Film Festival until 1990.

Then came the 1989 festival. It would change everything.

Steven Soderbergh's *sex, lies, and videotape* became the first festival film to emerge from obscurity and a relatively low budget ($1.2 million) into a mainstream hit. Shot on location in Baton Rouge, Louisiana, the film starred Andie MacDowell, James Spader, Peter Gallagher, and Laura San Giacomo as a quartet of people testing their erotic impulses and inhibitions. The film won the newly created Audience Award and then went on to win the Palme d'Or at Cannes.

Hollywood saw many reasons to take a deep dive into indie filmmaking, especially at Sundance. The studios were shouldering the burden of large creative expenditures in developing countless movie projects, many of which never went before the cameras, and then in making those films. What better way of filling out a release schedule than sending acquisition execs for a week to a ski resort to pick up a ready-made film or two?

Sundance also introduced them to talent they might otherwise be unaware of. Whether you picked up a newcomer's film or not, you could get into business with him or her. Thus, Sundance was able to bridge the historic fissure between art and commerce.

***

So where are we now in our search for any meaning in the term "independent film"?

Well, a truly "independent" film involves two different concepts. One is based on how the film gets financed. The other focuses on the spirit or vision of the filmmaker. If you raise money on your credit cards or with family and friends, or do deals with European television or (in those days) for video rights, you're an independent. Yet some, such as Alexander Payne, rely on a personal vision and innovative spirit to achieve the status of an independent. His films don't follow the tight storytelling of conventional Hollywood filmmaking, and with actors like Laura Dern and Reese Witherspoon, he can let things play out to see where this takes the movie.

In her book *A Killer Life*, indie producer Christine Vachon, founder and president of Killer Films, makes this point about independent films: their lower budgets mean a studio's risk is also lower. "Which allows me to push for the kind of independence in the filmmaking process that is crucial for our writer-directors. 'Independent film' as a media brand never interested me. And trust me, 'independently financing' a film only makes my job harder. But guarding a filmmaker's autonomy and agency—to tell unconventional stories, to cast the right actress not the star, to reject studio notes, to cut a third out of the movie right before the delivery date—is everything, since those values are what make film an art form and not just entertainment."

So this is the environment out of which the major players of *Sideways* emerged. Rex Pickett, an indie filmmaker if there ever was one, with two road movies to his credit, wrote the novel that kicked the whole thing off. Writer-director Alexander Payne's first film, *Citizen Ruth*, made its debut at Sundance in January 1996. *Sideways* producer Michael London, then a studio suit himself but looking for a way to make films he felt passionate about, was in the cheering audience. And most certainly Paul

Giamatti, whose lead role in *American Splendor* took 2003 Sundance by storm, with the film winning the Grand Jury Prize for Dramatic Film, was often an indie guy.

It's not a tremendous leap, as this is a book about film and wine, to equate our newly defined independent filmmakers with the mavericks in the wine world that stuck boldly to Pinot Noir in the face of relentless rejection from the public, some wine critics, the banks, and the trends and fads that plague the world of wine just as they do the world of cinema.

When Gary Pisoni, a Pinot fanatic, wanted to plant the family farm in the Santa Lucia Highlands of California's Monterey County with his favorite grape in the early 1980s, having done enough research to believe Pinot was well suited to that landscape and climate, the banks refused his plan.

"If you went to borrow money from a bank, they looked up [the] *California Grape Crush Report* [a publication of US Department of Agriculture] and saw what grapes sold for the highest prices," recalls his son and now vineyard manager Mark Pisoni. "Cabernet does. So the bank says, 'We'll lend money but only if you plant Cabernet.' Cab doesn't grow here, but the bank guys didn't know that back then.

"My father was all in on Pinot Noir but no one wanted it. He took a huge risk."

This was in the mid-1980s, and Gary struggled. Today his Pinots receive rave reviews, and his grapes are highly prized by buyers. He is now one of the rock stars of California Pinot.

People self-financing films, like Bobby Roth, or making deals with distribution companies, like Alexander, took the same risks. These risks don't always pay off: witness Rex Pickett's disastrous track record as an indie writer-director. So *Sideways* arose out of risk-taking: it was a film about characters studios were reluctant to feature, with actors only one company would sanction, about a wine few if any other than wine geeks understood, and about a region off nearly everyone's radar.

2

Ode to Pinot Noir

What is it about Pinot Noir that drives the movie's protagonist to such euphoric heights? Why does this one grape so enchant him with its mysterious allure? Why are so many adjectives deployed by critics to explain the grape's charms—long, intense, floating, sinewy, mineral, supple, silky, exotic, luscious, opulent, spicy, concentrated, ephemeral, delicate—and so many flavors and smells used to describe its tastes and aromas—cherries, plums, rhubarb, pomegranate, strawberry jam, mushrooms, leather, black pepper, and giving the French the last word, *sous bois*, meaning forest floor?

Pinot Noir is finicky. It's fussy like a spoiled child unhappy with her Christmas gifts, persnickety in the vineyard, and temperamental in the cellar. It seduces, teases, infuriates, spurns, and sometimes transports a wine enthusiast to the heights of ecstasy. Of all the grapes we use to make wine in America, meaning mostly *Vitis vinifera*—the European species used for wine grapes and brought to the New World by European settlers—Pinot Noir stands aloof and apart, sometimes eluding the best efforts of winemakers and creating despair among its millions of lovers, who, having once tasted its nectar, pursue the ethereal liquid sunshine in glass after glass and bottle after bottle.

The most expensive bottles of wine in the world, which come from Romanée-Conti in the Domaine de la Romanée-Conti in Burgundy, are the epitome of Pinot Noir. A single bottle might fetch anywhere from $4,000 to $8,000 and up—depending on vintage *and* whether you get a bargain. Older bottles might get north of $30,000.

In the opening chapter of his novel *Sideways*, author Rex Pickett has his alter ego, Miles Raymond, say this about Pinot: "*My* grape. The one varietal that truly enchants me, both stills and steals my heart with its elusive loveliness and false promises of transcendence. I loved her, and I would continue to follow her siren call until my wallet—or liver, whichever came first—gave out."

An ever so slightly different Miles, this one conjured forth by the film's writers, Alexander Payne and Jim Taylor, elaborates in a key scene in the movie. Here, where the gentle probing and queries between two lonely people during a night of temptations contain double meanings, Maya, the oenophile and waitress, asks Miles why he is so "into Pinot."

Miles considers this question for a moment, then says, "I don't know. . . . I don't know. It's a hard grape to grow, as you know, right? It's thin-skinned, temperamental, ripens early. It's, you know, it's not a survivor like Cabernet, which can just grow anywhere and thrive even when it's neglected. Yeah, Pinot needs constant care and attention and, you know, in fact, it can only grow in these really specifically tucked away corners of the world. And . . . and only the most patient and nurturing of growers can do it, really. You need somebody who really takes the time to understand Pinot's potential [who] can then coax it into its fullest expression. And, oh, its flavors, they're just the most haunting and brilliant and thrilling and subtle and ancient on the planet."

No less than Richard Sanford, the man who planted the first Pinot in Santa Barbara, says that Miles's speech "is the most beautiful expression of Pinot Noir I've heard given. It gave me chills."

Fellow Sta. Rita Hills Pinot vintner Bryan Babcock (Babcock Winery) concurs: "Miles's speech is perhaps the sexiest speech in the history of wine."

Wine authority Karen MacNeil notes that a computer search of terms used to describe Pinot would reveal that, more than any other variety, Pinot is described in sensual terms. This, she says, is because of "the remarkably supple, silky textures and erotically earthly aromas that great Pinot Noirs display."

"It's sexy to say in a restaurant, easy to drink, and there's more than great flavor and taste—there's a tactile sensation," says Adam LaZarre, a Paso Robles–based winemaker who's been making Pinot for three decades. "When you get a great Pinot Noir, it's like the angels are singing."

***

At its best, Pinot Noir is a seductive and smooth wine with an elegant nose of red berry fruits and layers of warm earthy notes. It is at once graceful and earthy, musty and fresh, pungent and fragrant. From top Burgundies, you can often pick up hints of what the connoisseurs call "barnyard" flavors or simply a "gaminess," which sounds off-putting but has inordinate appeal. When it's not at its best? How about green, bitter, weedy, thin, or flat? Little wonder consumers must beware. And little wonder winegrowers often despair.

It has more than once been called the "heartbreak grape," most notably by journalist Marq de Villiers in his 1993 book about California's pioneering Pinot producer Josh Jensen (of Calera fame) entitled *The Heartbreak Grape*. Along with the thin skin and fragility that Miles mentions, it can mutate rather easily, so a grower doesn't always know if the vine he plants will produce the wine he imagines. That thin skin also makes the grape susceptible to disease, rot, and pests. It demands specific well-drained soil and cool temperatures in those "little tucked away corners of the world," usually with limestone and marl soils. Those temperatures give Pinot a longer stay in the vineyard than, say, Cabernet, because Pinot develops its flavors by hanging a long time on the vine.

At harvest, it must be handled ever so gently so as not to break the skins prematurely and thus oxidize the juice. A winemaker must love this grape to devote so much time and energy to its planting, growth, harvesting, and vinification. Even then you can expect disappointment.

"Pinot Noir is hard to grow," states Santa Barbara Pinot grower and winemaker Wes Hagen. "It's the princess in 'The Princess and the Pea.' You can lay her down anywhere, but if she's uncomfortable, she'll let you know pretty damn quick."

Also, as Rajat Parr, a world-renowned sommelier turned winemaker, notes, "It's expensive to farm Pinot Noir. It's hard to machine pick because of its delicate skin. It's hard to have massive yields every year. It's more difficult to grow than other varieties. It's disease-prone, so you have to stay on top of it."

"Pinot is notorious for ripping through fermentation," sighs Kyle Altomare, assistant winemaker at Artesa Vineyards & Winery in the Los Carneros region of Napa Valley. "It's one of the quickest fermented varietals I've ever worked with. We've had wines that take off on their own and they're dry overnight."

He sighs. "You have to love Pinot. It's a cruel, cruel varietal, the cruelest I've ever worked with. You cannot take anything for granted. You really have to pay attention, and that's what makes it so elegant, so beautiful."

Perhaps this is why the great André Tchelistcheff, the father of modern California winegrowing and the man who helped define the style of its number-one red grape, Cabernet Sauvignon, notoriously mused, "God made Cabernet Sauvignon whereas the devil made Pinot Noir."

The latter sentiment can be read two ways: that the grape is hell to vinify or that Pinot may let you love her but she deceptively slips away after that first encounter, leaving only an unexciting version of herself in the next glass.

Mark Tchelistcheff, his grandnephew and a Berlin-based filmmaker who made a movie about his famous relative, *André: The Voice of Wine* (2017), happily confirms that André loved the grape beyond measure and that it was indeed his favorite wine. He was even buried in Napa Valley, the heart of Cab country, with a bottle of 1946 Beaulieu Vineyards Beaumont Pinot Noir, his favorite of all the wines he made.

"André would say that if you pick a fine ripe rose just before it has begun to open, take it home, put it in a vase without water, and the next morning put your nose into the heart of that dying rose and inhale its unique perfume, then you will have that aroma and scent to guide you to the perfect Pinot Noir."

One of André's star pupils and, arguably, the dean of California Pinot Noir, Greg La Follette, who is as much an educator as a great winemaker, explains how hard he tried to avoid having anything to do with the variety: "Pinot Noir is a really difficult varietal chemically. When people say it's a wimpy grape, it's because chemically it actually is. What it boils down to is in almost all varietals most color molecules are more robust, whereas Pinot Noir color molecules are not because they're nonacylated. So I knew having studied Pinot Noir there was no way in hell that I was going to make Pinot Noir because I was not an idiot. They don't call it the heartbreak grape for nothing.

"A lot of how you make Pinot Noir had not been worked out here in the New World. I knew I was not going to do Pinot because I was not a sadomasochist. But everywhere I went around the world I kept getting pulled, almost kicking and screaming, back into the world of Pinot Noir until finally I surrendered to Pinot. I said, 'Take me—I'm yours.' Ever since then, it's been easier. I spend a lot less money on therapy now that I surrendered to it."

***

As Greg says, the New World had much to learn about making Pinot. In the 1960s and 1970s, California Pinots often earned bad reviews and consumer dissatisfaction that drove many wineries out of the Pinot sweepstakes. The trouble was California's then cookie-cutter winemaking was applied indiscriminately to all red varieties. Red wine was being vigorously pumped, processed, and seasoned. Cab and Zin tolerated this treatment reasonably well. Pinot, with weaker tannins and thin skin, tended to lose color, display off flavors, and even oxidize. To rescue such wines, vintners would blend in darkly colored and deeply flavored varieties such as Carignane, Zinfandel, and Petite Sirah. Ugh.

There were exceptions. Among the pioneers of Pinot, Hanzell Vineyards in the foothills of Sonoma's Mayacamas Range focused on Pinot right from the get-go, the original blocks planted in 1953 and 1957. It had an owner, James D. Zellerbach, a San Francisco–based paper products magnate and former ambassador to Italy, who was rich enough and determined enough to produce stellar Burgundian varieties.

Meanwhile, Joseph Swan's 1973 in Sonoma's Russian River Valley and Eyrie Vineyards' 1975 South Block Reserve in Oregon's Dundee Hills kept the home fires burning until others figured it out.

Then there was Josh Jensen, founder of Calera Wine Company (established in 1974), who is hailed as the father of American Pinot Noir. In search of limestone rock and against all odds he pioneered the remote and rugged three-thousand-foot peak in the Gavilan Mountains that divide Monterey and San Benito counties. He not only crafted the Calera clone but championed and established his own Mt. Harlan AVA in San Benito County.

Eventually, winemakers realized that gentle handling, minimal manipulation, and a noninterventionist mindset would yield excellent results if the vineyard site was right. Still, Pinot can get moody in the cellar, going in and out of phases (also later in the bottle, to everyone's frustration), which causes some winemakers to intervene against their better judgment.

Then, too, Pinot Noir is a transparent wine. It may show flaws that other tougher varieties can hide.

"Pinot Noir shows more craft—mistakes or successes—than any red variety," insists Wes. "Pinot Noir suffers no fools."

Adam says the Pinot winemaker must be determined to find great Pinot vineyards to work with, because "you can't make it any better than the quality of the fruit that comes in. You can't manipulate it. You can only make it work or screw it up. You need to find one of those vineyards where the wines make themselves."

And that great vineyard will speak to the drinker. "Pinot Noir carries its sense of place better than any other varietal," says Siduri winemaker Matt Revelette.

Pinot historian John Winthrop Haeger makes the point that the second-generation pioneers who rejected the old red-wine recipe of wine technology often operated on shoestring budgets, much like the indie filmmakers we talked about earlier. Operating sometimes without electricity and with home-manufactured ersatz plungers to punch down the cap of skins or simply stomping with their feet, they *removed* technology from the cellar and followed a recipe conceived specifically for Pinot Noir. Then when this new recipe was later combined with better fruit, farmed to lower yields, and picked later, Pinot emerged as varietally correct and a more nuanced and complex wine.

Pinot Noir from Sonoma Mountain, Sonoma County, and Los Carneros, Napa Valley, California AUTHOR'S COLLECTION

Pinot Noir, Santa Maria Valley, Santa Barbara County, California AUTHOR'S COLLECTION

Winemakers Greg La Follette and Evan Damiano at their Marchelle winery, Sebastopol, Sonoma County, California AUTHOR'S COLLECTION

Pinot Noir is also just about the only red grape that never gets blended with other grapes. There is no excuse for anyone to blend Pinot with anything else. That would miss the whole point.

"I've tried to blend everything I can think of into Pinot Noir," admits Adam. "I've tried and tried and it's really hard to find anything you can blend into Pinot. Maybe Chardonnay, an old Burgundy trick, to brighten the fruit with 2 to 3 percent. Anything else completely wipes out the Pinot Noir character. It doesn't taste like Pinot Noir."

***

Pinot Noir is an OG—an original gangster, meaning an originator, authentic, old school, been around forever. No doubt thousands of years old, it belongs among the "founder varieties" as an ancestor to scores of other grapes. All Pinot Noir comes from Burgundy.

The wine grape varieties of the world are believed to be descended from a single plant species, identified botanically as *Vitis vinifera* although some now extinct species may lurk in the pedigree of some modern varieties. The varieties of *vinifera* are largely—possibly entirely—the product of cultivation. In other words, man intervened sometime between 10,000 and 7000 BCE. Joe Cave Dweller picked a lot of grapes one warm summer's day, ate some, put those left in whatever passed for a jar in that era, forgot about it, then discovered the now fermented grapes, and in drinking the juice experienced the thrill Miles Raymond gets quaffing a Highliner at The Hitching Post.

The University of California–Davis and especially Carole Meredith, its now retired viticulturist and plant geneticist, have used DNA fingerprinting—the very technology used in every cop show on TV—to identify the parents of our modern wine grape varieties. Here's the thing: no parents have been found for Pinot Noir. None. This leaves open the possibility that the parents may have been wild vines. The romantics among us Pinot lovers certainly hope so!

What Meredith's DNA-based work does reveal is that Pinot Noir is one of the genetic parents of Chardonnay, Gamay, Aligoté, and at least thirteen other varieties. Because of its old age and genetic instability, Pinot Noir has thrown off hundreds of clones. Tasting room personnel and winemakers themselves will regale you to no end with the clonal selection that went into that glass of Pinot in your hand—Dijon, Martini, Wädenswil, Pommard, Swan, Calera, and so on.

This all goes back to the greatest disaster to ever hit the storied vineyards of Europe. In the late nineteenth century, phylloxera, a devastating insect that destroys vine roots, befell vineyards all over the continent and drastically changed the way growers plant and cultivate grapevines. Sometime in about the midcentury someone innocently planted a vine sent by a friend or relative from America on French soil. Clinging to that root were countless nearly invisible yellow insects. These were harmless and unknown in the United States because indigenous vines were tolerant of the pest. European vines, *Vitis vinifera*, were not. This was a calamity. The end of the world's vineyards grew near.

Scientists finally hit upon a solution. It went against the grain in more ways than one, but vineyards around the globe eventually grafted most of their vines onto American rootstock. So yes, that Burgundy or German Riesling you're swirling in a glass has a bit of Yankee blood in it.

The mass replanting led to a demand for reliability and consistency, which spurred vine breeding and propagation programs throughout Europe. Even a virus-free vine, however, might be unpredictable, with variable yields or uneven ripening. So researchers took cuttings from vines with specific and desirable traits. They'd graft them, plant them, and watch to see if they carried the desired traits. If so, they'd propagate cuttings from those vines again through several generations. All the cuttings could therefore be traced directly to the initial mother vine, and all shared the same DNA. Hallelujah! Clonal selection was born. Germany, not France, led the way in this science.

California began the same process even earlier, when Dr. Harold Omo of the University of California–Davis began to import Pinot vines from France, Germany, and Switzerland. He also selected vines from pre-Prohibition vineyards. His work would eventually result in UC Davis's Foundation Plant Services (FPS), founded in 1958.

As FPS evolved, methods to heat-treat cuttings developed to make the material virus-free. But as Larry Hyde, one of the pioneers in Pinot plantings in Napa's Carneros district, points out, this heat treatment "changed the structure" of the vines. New technology has

Pinot Noir from Los Carneros, Napa Valley, California AUTHOR'S COLLECTION

evolved, though, so that "now if you take plant material to UC Davis they have a new way to treat it so it doesn't change the genetic structure."

Does any of this matter?

Clones are a major viticultural achievement, play a significant role in modern winemaking, and will help meet the challenges ahead due to climate change. The same Pinot clone planted in different vineyards can look distinctly different, and even the taste may be different. The clones are the same but the expression is different.

Katie Santora, the winemaker at Oregon's Chehalem Winery, likes to focus her Pinot bottlings on the winery's single vineyards rather than clones. Yet she admits she uses clones "as building blocks to create complexity."

Ultimately, clones are one part of the vast matrix of elements that add to Pinot Noir's allure. Clones are different, and you can taste those differences in the flavors and structure of that Pinot in your glass. But they do not trump terroir, the French word for the total impact of any given site on the wine itself. That's the key.

A vigneron in Paso Robles, California, an area making very little Pinot because the climate is not hospitable to the grape, uses the term "magical chaos" to explain the variables involved in a bottle of his Pinot, Windward. Marc Goldberg ticks off these

Pinot Noir, Paso Robles, San Luis Obispo County, California © ZW IMAGES

influences: the countless nutrients available in his soil, the effects of the Templeton Gap's cold winds blowing through the region, four Pinot clones, and his own influence.

"The vineyard makes the wine. That's why I don't call myself a winemaker. I'm the wine shepherd. I follow the grapes in and I protect my crop from bad things that can happen to it, dirt and harmful bacteria. So you're not drinking *my* wine; you're drinking my vineyard."

***

So Pinot may have arrived in France with the Romans or the Greeks or was domesticated in situ from indigenous *Vitis vinifera*. The point is it got there. Burgundy's Valois dukes, whose duchy extended from the Alps to Flanders and whose power rivaled that of the kings of France, promoted the Pinot wine—"the finest in the world"—with a propagandistic zeal that would put a modern-day marketer to shame. The variety was transported northward to Champagne, west to the nearby Loire, east to Alsace and Germany, and southward across Switzerland to Italy.

In the nineteenth century and with great passion in the twentieth century, Pinot was transported and planted in various parts of the world, including South Africa, Chile, Brazil, Argentina, Australia, New Zealand, and of course North America. The top three Pinot-producing countries are France, the United States, and Germany.

The wine can be unstable. It can suffer from bottle shock. It can be moody and capricious. One of the authors remembers an ethereal 1981 Clos de Vougeot, sampled over several bottles over several months. Upon drinking the final bottle, the wine had turned dull and flat. Who knows why?

Pinot may be a victim of its legendary status. Jim Clendenen complained that consumers and critics reserve a special onus for Pinot by demanding that *all* Pinots achieve greatness. They permit Cabernet, Syrah, and Zinfandel to produce respectable and serviceable wines. Yet Pinot is held to a higher standard.

"Burgundians have lived long enough with pinot noir to have become accustomed to its way and to accept its foibles as the price of sometimes stunning wines," notes John Winthrop Haeger. "But on this continent, aspiring winemakers and unhappy critics have embraced an odd vocabulary redolent of epic strife and operatic drama without which, it seems, no story about Pinot Noir is ever complete."

One must accept that uncertainty is inextricable from the Pinot Noir experience. Vintages vary, especially in the Old World, with wildly varying climates from year to year. Different growers with adjacent parcels in the same vineyard or a nearby vineyard will make different-tasting wine. There remain fibers of consistency running through such wines, but to appreciate Pinot Noir you need a tolerance for surprise.

3

2 Guys on Wine

It was, he later recalls, a "brutal time." His indie filmmaking career—stalled. His marriage—collapsed. His mother, a year and a half after her husband's death, was overwhelmed by a massive stroke. When he returned from a failed writing assignment in England, he discovered much of his mother's money had disappeared into his brother's pocket. Then his agent died of AIDS. Forced to sublease his rent-controlled house in Santa Monica, California, which was illegal, he was essentially broke. Yet these "roommates" drove him nuts.

What saved him was golf.

When Rex Pickett was a young teenager in San Diego, he was a promising enough golfer that he ranked number two in his age group in the San Diego Junior Golf Association. Now, in his thirties, he returned to golf with renewed passion. He needed "something to get out of my head." For his head was not in a good place to be in those days.

Rex had a two handicap and a taste for rough courses. But Los Angeles was not the place for a destitute golfer who was certainly no member of an elite country club. That's when he discovered a golfer's dream.

Located on 309 acres of oak-studded ancient sand hills, La Purisima was built in a time of boom for the city of Lompoc. The sleepy Central Coast town had grown slowly until the midcentury establishment of Vandenberg Air Force Base, the Air Force's first missile base. Then in the 1980s, the space shuttle program announced it was going to begin launches from that site. Restaurants, hotels, condos, and a golf course sprang up in anticipation of the huge tourist attraction of frequent shuttle launches.

Then the Challenger exploded during takeoff from Cape Canaveral in 1986. The West Coast shuttle program was terminated, sending the town into a deep recession. But the golf course, La Purisima, remained.

What Rex found when he walked onto the golf greens was an extraordinary course, uncrowded, with no homes lining the fairways and sublime natural beauty.

Initially, he drove up and back in a day from Santa Monica for a round of golf. But he thought, why not stay overnight to get in two rounds? The Windmill Inn in Buellton just off the 101 freeway was $29.99 a night. For dinner he could stroll across the freeway overpass, head east on highway 246, and eat at A. J. Spurs. Then he discovered further up the road another eatery called The Hitching Post. The latter destination would change his life.

***

Though born at Castle Air Force Base in Merced County in Northern California, Rex grew up in San Diego. His father, an Air Force captain, got drunk one day and punched out a major colonel. As a consequence, his dad found himself out of the service with three sons to support.

"Everyone in my family struggled with alcohol, each in their own discrete way," Rex now says.

He arrived in San Diego at age five, with two brothers and a mother who was a registered nurse. She got a job with a doctor, and his dad eventually found prosperity in the laundromat business, owning five at one point. He wanted Rex to take over that business, but his son preferred to build surfboards. He got accepted at UCSD, where he studied contemporary literary and film criticism and creative writing. This is where he encountered Manny Farber.

"Manny Farber is one of the very few movie critics who have mattered in this country," Richard Schickel, himself a film critic, biographer, and documentarian, once wrote. "Certainly, he was the first to think seriously and coherently about the American action film, thereby creating an aesthetic that allowed us to fully apprehend, for the first time, our native genius for moviemaking."

Farber wrote intensely personal essays following trains of thought down many a rabbit hole in which a reader either experiences vertigo or joyfully gives in to the perceptions of a restless, quirky, densely read, and highly educated mind trying to link movies to the larger context of modern art.

By the time Rex took Farber's famous UCSD film classes, Farber had left New York and given up criticism to focus on teaching and painting.

"I took every class for the next five years. It was life transforming. It was on a Wednesday night for three hours with 350 people. It was an event. He rented films he wanted to see and didn't care if the students liked them. All the Europeans—Godard, Fassbinder, Wenders, Herzog, the French—New York experimental cinema. He asked me, an undergraduate, to be his teaching assistant, which never happened before."

The film that made the deepest impression on the budding filmmaker was Wim Wenders's *Kings of the Road* (1976), a German three-hour black-and-white road movie with almost no dialogue. "I thought I'm just going to make road movies," says Rex. "That's what I want to do with my life."

Now married to Barbara Schock, a fellow Manny Farber acolyte, he and his new wife plotted their first movie. A $100,000 trust fund was released to his bride the day they married, so much of that went into a one-thousand-mile road movie, *California Without End* (1984), shot in 16 mm. He was the writer, director, and editor, and she was the female star.

"It played festivals—kind of a bleak film," he admits. *California* sold to Bavarian Radio Television, a German TV station.

During this frenetic, energizing, and wearying time of making an indie film without asking permission from anyone, the couple entertained a couple of visitors at their Santa Monica house on Twelfth Street. A girlfriend of Barbara's came down to visit from Stanford University with her boyfriend. The young man was amazed by the creative environment and passion for art he found in their home. His name was Michael London and he would one day be the producer of *Sideways*.

"They were passionate, passionate artists and had great dreams of the future and the movies they wanted to make," Michael remembers. "Rex was the first true artistic filmmaker I'd ever met—someone really following his dreams about stories he wanted to tell and movies he wanted to make without any connection to the business. This was a moment when there wasn't a tradition of American indie filmmakers. But he was determined to follow the path of the great European filmmakers."

After their first film, the couple hit the road again. This time the road trip went *From Hollywood to Deadwood* (1988). Barbara, who would produce the film, raised the $500,000 budget through a limited partnership. The movie, shot in Super 16, concerned two sorry detectives on the trail of a missing starlet. The road picture more than quadrupled down on the previous film's mileage, as it ranged four thousand miles starting in San Diego, roaming to Los Angeles, then all the way to the Black Hills of South Dakota, and back to LA.

"That took five years out of my life," sighs its writer-director. "We had problems because we were undercapitalized."

When Island Pictures bought the film for $650,000, $100,000 was earmarked for shooting R-rated scenes for the foreign market.

The film languished in postproduction for two years. Following a screening at the Mill Valley Film Festival, about 20 percent of the footage was cut by the president of Island Pictures. The film's investors broke even. Rex got nothing for writing, directing, and editing.

But wait—a review in *The Hollywood Reporter* out of Mill Valley caught the attention of Kevin Bacon's MixedBreed Films, based at Columbia/TriStar. The company was looking for someone to write a mystery. Rex wisely sent his original script and not the now-butchered movie to MixedBreed Films.

Bacon's partners liked the screenplay. Someone called Rex and told him he was their writer. "For the first time, I get paid as a writer. I get validated. Yes, I made two pictures but they were financed by my wife. They still felt like vanity projects."

***

That long-ago visitor to the couple's Santa Monica digs, Michael London, wound up at 20th Century Fox as a vice president of film production. He was now in a position to help get Barbara Schock a job.

While at Stanford, Michael had interned at *The Los Angeles Times*. When he came to LA following graduation, his former editor invited him back as an editorial assistant for Calendar, the paper's entertainment section. He knew little about TV or film. He would only write about such things because that was what the job demanded.

So at a fairly young age, he soon became a *Times* staff writer. In a few years, though, he realized journalism was not his path. "Daily journalism is a bit of a grind," he says. Then came a serendipitous job offer.

A profile he wrote about the artistically adventurous filmmakers the Coen brothers, Ethan and Joel, caught the attention of somebody at Don Simpson/Jerry Bruckheimer Films on the Paramount lot. Somehow that piece of journalism got him a one-year contract in development with two of the hottest producers in town, the talented and steadfast Jerry Bruckheimer and his bad-boy partner Don Simpson, the men behind such 1980s superhits as *Flashdance, Beverly Hills Cop*, and *Top Gun*.

This led to a studio job as an executive at Fox. And to Michael being able to lend a helping hand to his old friends.

Schock came aboard the Fox lot in 1990 as personal assistant to director Vincent Ward, who was trying to figure out how to squeeze an *Alien*[3] out of a tiring space-horror franchise. When Ward came off the picture, the studio took the unusual step of hiring a first-time director named David Fincher, twenty-five, who had previously done music videos. Schock then became his assistant.

The production was already bivouacked in Pinewood Studios outside London. Fincher was unhappy with the script he inherited from Ward, written by Walter Hill and David Giler, who were the official producers of record for all the *Alien* movies. Schock called her husband, the screenwriter.

"With Barbara liaising between her and Fincher in London and me in Santa Monica, I rewrote scenes from the Walter Hill/David Giler script," says Rex. "Fincher liked my work. On his own dime, he flew me to London, where I worked for free under the cloak of darkness. In nine days, I rewrote the entire Hill/Giler script with only Fincher and Barbara knowing. Then Fincher did something incredible. He presented this to Joe Roth [then the studio's head]."

Rex says he then became a victim of Hollywood's power politics. "Walter Hill threatened to pull [the film's star] Sigourney Weaver. My version was pulled at Hill's order from all departments and shredded. Barbara and I were fired."

***

In the middle of the *Alien*[3] fiasco, Rex got a call in London. His mother had suffered a massive stroke and was in the ICU. But he was trapped by the chaotic preproduction and couldn't leave. When he returned home, he discovered his brother had misappropriated much of his mother's nest egg.

He got his mother into an assisted living facility. He assumed control of her health care from his wayward brother and sold her condo for "very little."

In 1990, Schock got a job as assistant to Michael Nozick, the executive producer of *Thunderheart* (1992), a Michael Apted crime thriller. The two got along well, so when Joan Micklin Silver, one of the original indie filmmakers in America, was looking for a head of development for her Silver Films Productions in New York City, Nozick recommended her.

Schock moved to New York. She didn't serve her husband with divorce papers until 1995, but their marriage was over. Which didn't prevent her from passing along a spec script written by him, *The Road Back*, to her new boss. It concerned a music video director whose mother suffers a stroke, much of this related to his experiences tending to his mom. Silver Films optioned the script.

Rex remembers Joanna Woodward at one point circled the wheelchair role in *The Road Back* before bowing out. So it was never made, but that story would ultimately form the backbone of Rex's second *Sideways* novel, *Vertical*.

Which was how he ended up back in his rent-controlled $900-a-month house on Twelfth Street in Santa Monica, his agent now deceased, illegal roommates getting the better of him, and his filmmaking career in tatters.

He continued to write, which was all he knew how to do. And then there was golf.

When he ate dinner at The Hitching Post, he discovered the bar off to the right of the entry waiting room. The joint was touristy on the weekends, but on weeknights, winemakers hung out, and in the cozy family restaurant to the left of the entry they would take wine writers to dinner to sample their wares.

Its owner, Frank Ostini, a restaurateur and winemaker himself, recalls, "Rex found out about the region being a wine region by coming to our restaurant. He befriended a winemaker at our bar, Chris Whitcraft, and found out about our [region's] wines."

Wine was a thing Rex had never taken seriously. But he was suddenly meeting people who did.

Living down and out in Santa Monica, he soon gravitated to a wine store on Montana and Sixth called Epicurus. It used to be a Blockbuster Video, a long and narrow affair. On Saturdays from three to five in the afternoon, wine reps would come to pour their wines for free. Which suited the nearly broke ex-filmmaker just fine. Imbibers would congregate in a small cordoned-off tasting area in the rear affectionately named The Bullpen.

"It was my only social outlet," he says.

More importantly, he befriended a British guy named Julian Davies, who worked there. He knew his wine from A to Z, so when the store's owner left early to rendezvous with his mistress, Davies would take revenge on his low salary by opening prize bottles and sharing the legend and lore of wine with a guy new to its wonders.

Julian would uncork bottle after bottle as the oenophiles swirled and sniffed, angled wineglasses at forty-five degrees against white backdrops, sipped, then swallowed, and finally breathed out. Everyone was searching for the flavors and aromas of the wines. What was the fruit like? Was that coffee or chocolate? Earthy, tobacco, oaky? They detected odd smells, too, say petroleum in Rieslings and cat piss in Pinot Grigio. What fascinating people these wine folks were.

Julian introduced the struggling writer to the glories of Pinot Noir from Rochioli, Gary Farrell from Sonoma, and Whitcraft and Au Bon Climat from Santa Barbara. Rex learned about Williams Selyem's single-vineyard wines and Comte Armand Pommard from Burgundy.

He started reading about Pinot Noir and Burgundy in Jancis Robinson's wonderfully comprehensive *The Oxford Companion to Wine*. Then he discovered The Hitching Post.

"This is heaven," says Rex. "The wineries along the Foxen Canyon Trail and the town of Los Olivos where there were three tasting rooms and now there's fifty. So I'm playing golf, tasting wine in the Santa Ynez Valley and Santa Monica, and I decided to write a mystery novel."

He called it *La Purisima*.

He had no agent, but for once a man who had seemingly run out of luck found some. An assistant to his former wife in New York got the *La Purisima* manuscript to a lit agent at Curtis Brown, Ltd., named Jess Taylor. The man liked the novel, called Rex, and told him it needed work but he would like to represent the novel.

"So I get a little bit of life," Rex says. "Someone blowing on the embers."

After Rex did two more drafts, Taylor went out with the manuscript. Then came the rejection letters, not unlike those received by Miles in the eventual movie *Sideways*.

Still, the validation of having a New York agent buoyed the spirits of the nascent novelist. And in a strange turn of events, he acquired a second agent. Jess Taylor got lured to Hollywood to become a book-to-film agent at Endeavor Talent Agency, which handled a large number of film and TV stars. (In 2009 it would merge with the storied William Morris Agency. Today the large talent and media agency is known simply as Endeavor.) Taylor passed his client and his novel *La Purisima* along to a new agent at Curtis Brown, and Rex found himself all of a sudden with a film agent at Endeavor.

He kept returning to The Hitching Post.

"Rex was a tortured guy," says Frank. "His reputation was . . . well, he'd walk from the Windmill Inn to have dinner here because he knew he'd be drunk and wanted to walk responsibly back to the hotel."

Along the Foxen Canyon Wine Trail, he discovered the superb Pinots crafted at Foxen Winery, cofounded by two close friends, Dick Doré and Bill Wathen, in 1985. Many years later, the two would build a solar-powered winery and tasting room down the road, but in those days, tastings happened in a dilapidated shed along the Trail that looked almost abandoned. Jammed into this crowded space were a small bar counter and a funky shrine loaded with odds and ends, photographs, memorabilia, and cartoon strips collected over the years. (In *Sideways*, it can be seen in the tasting room where Thomas grabs a bottle and pours more wine into the naughty boys' glasses when their server disappears for a moment.)

Rex quickly made friends with Dick. There was also a female tasting room manager Rex found quite attractive.

"He'd come to our winery and sit there all afternoon and lust after Sandy—he was in love with her," says Dick.

Dick Doré learned to tolerate his new friend's bottomless appreciation for his wine.

"We had a wine event at the winery once and he's drinking all day," Dick recollects. "Then about eleven, he said, 'I'm ready to go.' Okay, so went back to our house

Owner/winemaker Frank Ostini at his Hitching Post Wines' tasting room in Buellton, California. In the background photo, Ostini with his partner/cellar master Gray Hartley AUTHOR'S COLLECTION

The Hitching Post bar counter, Buellton, California AUTHOR'S COLLECTION

[just across the road]. We got there and he said, 'Okay, what are we going to drink now?' So I went down to our cellar and we drank a bottle. 'Well, what are we going to drink now?' he said. Jenny [Dick's wife] went to bed. We sat up till three o'clock in the morning and we had four bottles of wine. He likes his wine."

Rex began taking along a friend for his lost weekends in Santa Ynez. Roy Gittens, who was his gaffer, as chief electricians are called, on *From Hollywood to Deadwood*, was a charismatic guy, not quite the Jack character in *Sideways* but a handsome, funny guy. The two played golf and went wine tasting.

(Thomas Haden Church recalls a crew member approaching him on the *Spider-man 3* set and introducing himself as the model for his character in *Sideways*.)

Dick Doré and Bill Wathen, founders of Foxen Vineyard & Winery, Santa Maria Valley, Santa Barbara County, California PHOTO BY KIRK IRWIN

Foxen's rustic beloved tasting room shack, Santa Maria Valley, Santa Barbara County, California
AUTHOR'S COLLECTION

Pinot Noir planting, Spanish Springs vineyard, San Luis Obispo Coast AVA, California. AUTHOR'S COLLECTION.

One afternoon as Rex and Roy were getting seriously looped on the wine trails, Gittens turned to his pal and said, "Rex, you should write a screenplay about all this."

Rex nodded sagely, as inebriates are prone to do. Sounded good to him.

Their next stop was the Fess Parker Winery. In the large tasting room with high ceilings, flagstone floors, comfortable sofas, and a blazing fireplace, Rex noticed the former TV frontier hero himself, sitting in a high director's chair wearing his trademark coonskin cap. After buying a bottle of Fess Parker Pinot, Rex approached him with the bottle and asked him to sign it.

Fess Parker graciously asked what the younger man did.

"I'm a writer," he replied.

"What do you write?" asked the actor.

"I write screenplays," he declared.

"What are you working on now?" asked Parker.

"Well, Fess, I just had a brainstorm," he said. "I'm going to write a script called *2 Guys on Wine*."

Parker smiled, then wrote, in a gold pen designed to write on bottles, "For 2 Guys on Wine. Best, Fess 10/1996."

***

Rex did what he told the former TV star he would do: he dashed off a screenplay entitled *2 Guys on Wine*.

"Frankly, I didn't like it," he says. "I thought it was puerile and I didn't show it to anyone."

With rejection letters piling up and creditors making threatening noises he tried to shut out, he had nothing better to do than embark on a short story. It was a first-person tale about a guy named Miles who goes to the wine tastings at Epicurus. In the story, the whole sniff-swirl-sip ritual turns into an out-and-out brawl. He rather liked that twist. When Rex got to the end of the story, a thought hit him.

Wait a minute, the screenplay I didn't like is actually a novel, Rex remembers saying to himself. This is the prologue. Jack comes in and saves Miles and they go off on this trip.

He began to write. As the writing continued he journeyed back to the Santa Ynez Valley. Sitting at The Hitching Post bar, there was someone in that restaurant he thought would be a good fictional fit.

"He had a crush on a waitress," Frank says. "They never dated but he wrote her into the book using a different name."

"One night, we got down over some Pinots after hours," Rex recalls. "Striking woman."

And, remember, he was also smitten by a tasting room manager back at Foxen.

So the novel was coming into focus. *2 Guys on Wine* would be Rex and Roy playing golf and getting looped in Santa Barbara tasting rooms. Throw in a Hitching Post waitress and a tasting room hottie. Then, like his two movies, this would be a story of a road trip, one that begins in Santa Monica, travels to the Santa Ynez area for various adventures and catastrophes, and finally heads further north to the Central Coast wine town of Paso Robles for a wedding.

The picaresque road trip is as old as *Don Quixote* four centuries ago and *The Odyssey* as many as 2,800 years ago. The road plays a big role not just in American literature but in movies as well: everything from Jack Kerouac to Cormac McCarthy via *Easy Rider* and *Thelma & Louise*. The road is such an ancient metaphor for life. This idea that you go down life's road, stuff happens, and you strive to be better. And you have some kind of goal, some kind of grail in mind.

Although his friend Roy was never about to get married during their carousing in North County, Rex invented the wedding to get a ticking clock into his story and a final destination, a grail, for a road trip. His Miles is more desperate and suicidal than the movie's Miles would be, gulping down almost as much bitter-tasting Xanax as wine for his anxiety. His despair is greater. Jack's various physical injuries during the week require *three* visits to Lompoc Hospital's emergency room, for a broken nose, broken ankle, and cracked ribs. Intuitively, Rex knew to seek comedy, albeit the painful kind, for such a story.

"If I write a novel about this lonely guy who's despairing and drinks too much, no one is going to buy it," he says. "So I need to find comedy. With comedy, you can sell despair."

Ever the student of Manny Farber, Rex had in the back of his mind a Satyajit Ray film he saw in his class. Ray is among the true masters of Indian cinema. The Bengali writer-director's restrained style and deceptive simplicity let him observe life, an environment, and its people with a clear eye for the follies and foibles of mankind.

In *Days and Nights in the Forest* (1970), he discerns the contrasts between urban and rural folks, men and women, innocence and corruption by sending four twentysomething professionals from Calcutta—now Kolkata—on a holiday in the countryside of northeast India. The four men have worked hard but found little relief in love or romance. They get drunk at a village bar and make fools of themselves, but they are not without sensitivity or good humor. Two beautiful women come within their orbit. As the six characters become familiar with each other, dynamics shift and tensions grow. This is not a movie of major incidents or crises but rather an investigation into the complexity of life.

Around 1998, Rex wrote a first draft and then a second. In September he gave the manuscript to Taylor and to Michael London, who by now had left Fox to pursue a career as an independent producer.

Everyone loved the novel. Taylor came over to Rex's house, bringing a bottle of Byron sparkling wine that Rex had given him as a gift when he came to Endeavor—the same bottle Jack will imprudently uncork in the car with Miles in the movie. Over dinner, Jess asked for minor changes to the opening and, more importantly, a title change.

"*2 Guys on Wine* sounds like a travelogue," Rex's agent told him. "Come up with a one-word title that's unique."

Rex found this in the early pages of his first chapter. The proprietor of Epicurus upbraids Miles for turning his wine tasting into marathon drinking: "If you didn't get so sideways on Fridays you might be on the last chapter of that novel of yours." From this point on Miles and Jack use the term frequently to mean inebriation.

Sideways.

The novel was dedicated to "Roy and Julian, partners in wine."

***

The plan was for the New York book agent and the Hollywood book-to-film agent to submit the manuscript to publishing houses and filmmakers simultaneously. Michael wanted to attach himself to the script but—and this is pivotal to a later disagreement—*without* paying an option. Taylor was wary of this, but Rex figured the more people on his side getting the unpublished manuscript out there the better. The manuscript then got to David Lonner, the agent for Alexander Payne, hot off *Election*. Lonner's office was down the hall from Taylor's because both men worked at Endeavor.

By the new year Rex's new lit agent at Curtis Brown, Mitchell Waters, pulled *Sideways* from submission. He was getting ruthlessly turned down by every publisher that read it.

Shaken, Rex showed the manuscript to his ex-wife. He says she told him to burn it.

A more serious blow came in March. At a book launch for one of his clients, Taylor sidled up to Rex and told him he was leaving the business. The ruthless world of agencies had destroyed him. After breakfast with him the next morning in Beverly Hills, Rex walked him to his psychiatrist's office. That was the last he saw of Taylor until the red-carpet premiere of *Sideways* at the Academy of Motion Picture Arts and Sciences in 2004.

4

Why Santa Barbara?

When the Spanish missionaries and king's soldiers first arrived in the New World, they planted around their homes essential vegetation—fig and olive trees, and grape cuttings brought from Spain. As these explorers moved further north from modern-day Mexico to the new territory of Alta (Upper) California, they seized political control from the native population by establishing a string of missions, each a day's journey from the next. Wine was essential for these Spaniards for sacramental ceremonies and for nourishment and relaxation in a harsh daily life.

Santa Barbara County's three missions—Santa Barbara, La Purísima Concepción, and Santa Inés—were among the twenty-one missions founded along the coast of Alta California between 1769 and 1823. The rootstock of *Vitis vinifera* was brought over by the famous Father Junipero Sera around 1782, along with knowledge of winemaking. For decades the so-called Mission grape produced a sweet, low-acid beverage for the settlers along with *aguardiente*—brandy—which was easier to make and preserve.

During the first half of the nineteenth century, settlers established many vineyards and small wineries in the Santa Barbara region. Then the Mexican-American War resulted in a Yankee land grab—the secession of Texas from Mexico and the birth of the California Republic, a breakaway state from Mexico, in 1846, a rebellion encouraged by US Army Brevet Captain John C. Frémont. Freed from Mexican control, California became the thirty-first state in the Union in 1850. Dick Doré is descended from one of these rebels, an English sea captain, William Benjamin Foxen. He was the fellow who, in 1846, warned Frémont (the Pathfinder) of an impending Mexican ambush down the road at Gaviota Pass. Foxen safely led Frémont and his men over San Marcos Pass, avoiding the ambush, and they captured Santa Barbara without bloodshed. However, his neighbors, Mexican families and Indians, felt betrayed by his actions.

"He felt he would rather cast his life with America than Mexico," Dick explains many decades later. "He got burned out and had to live at Santa Inéz Mission for two years. They tried to kill him."

***

Few people realize that Los Angeles was once the wine capital of the West. More than a million grapevines covered the ground. A French immigrant with the patronymic name of Louis Vignes, from a Bordeaux barrel-making family, established a commercial winery in 1833 near what is now LA's Union Station. He seems to have been the first to move beyond the Mission grape to plant more sophisticated wine cuttings from France.

Then an insect-borne bacterial infection called Anaheim disease crippled the Los Angeles wine industry in the 1880s. By then, thanks to the Gold Rush and escalation of the population of San Francisco, the California wine industry shifted north, first to Sonoma and then Napa, both valleys better suited to wine growing than Los Angeles.

In the town of Santa Barbara, Albert Packard built the first major winery, La Bodega, on West Carrillo Street sometime after 1865. In the latter half of the nineteenth century, forty-five or so vineyards of one to forty acres each dotted the countryside of Santa Barbara, Carpinteria, and Goleta and into the North County areas of Lompoc and Santa Ynez Valley. Some vintners even dared to experiment with wine grape varieties such as Zinfandel, Olivet, and Tokay.

The early California winemakers throughout the state were winging it. They had little schooling or formal training in viticulture, no traditions to build upon, and no idea which varieties would grow best in this climate or that soil. They struggled with crude equipment, and even bottle-making factories were scarce. Then phylloxera, an insect pest of grapevines worldwide, hit in the mid-1880s.

Yet nothing can deter the determined vintner.

The industry forged ahead in the state so that by the dawn of the twentieth century, California could boost about three hundred grape varieties and nearly eight hundred wineries, according to Karen MacNeil's *The Wine Bible*. Then came a calamity even determined vintners couldn't overcome. On the sixteenth day of January in 1920, the Volstead Act took effect.

Thus, America's love for wine and passion for wine growing and making suffered the greatest coitus interruptus in its wine history—Prohibition. A few wineries continued to make sacramental wines, and an intended loophole in the Volstead Act allowed home winemakers to make two hundred gallons of "medicinal" wines annually. Nevertheless, Prohibition's impact on American wine was catastrophic.

First of all, wine culture, which was starting to bloom in the United States, all but vanished other than among Italian and other immigrant families. If one thinks about alcohol consumption during this period, one thinks of speakeasies and bathtub gin and bootleggers—in other words, hard liquor.

Look at old movies from post–World War II into the 1960s and 1970s. Rarely do you see a wine bottle on a dinner table, as is common in films today. Wine disappeared from mainstream American culture, although in Santa Barbara as well as in many other parts of the nation a commercial wine tradition carried on at a neighborhood scale.

Then just as tragically, with all the wineries closing down due to Prohibition, that hard-earned knowledge of wine growing and making was lost. A robust nineteenth-century wine industry was wiped away. Few knew any longer how to make great wine in California, and not many across the country had any thirst for such a product. If swells on the East Coast wanted their claret, champagne, or Burgundy, they grabbed the real stuff from across the pond. A California Burgundy? Hmph!

One recalls James Thurber's classic 1937 *New Yorker* cartoon of a wine snob holding forth at a table over an open bottle. "It's a naive domestic Burgundy without any breeding, but I think you'll be amused by its presumption," he ruminates.

During America's postwar industrial boom, table wine was a sideline. The consumer thirst was for brandy and fortified wine. Gradually, however, UC Davis became the state's great academic engine for its wine industry, and a midcentury rebirth of California wine began, driven by such true believers in the Napa Valley as Robert Mondavi with his eponymous winery, John Daniel Jr. of Inglenook, and André Tchelistcheff of Beaulieu Vineyard, who crafted the era's great Cabernets, and Martin Ray on hills above Saratoga, who aimed to rival Burgundy with his Pinot Noir and Chardonnay.

Nevertheless, mistakes were made in Santa Barbara (and elsewhere). Santa Barbara vintners did a very poor job of grape growing in the 1960s and 1970s. Among other things, they attempted to force high yields of six tons an acre to recapture investments when the cool climate only permitted half that amount. Then they planted too much Cabernet, for most of these sites were too cold for the king of California reds.

What they eventually learned (relearned?) is that this land was made for the Burgundian varietals—Pinot Noir and Chardonnay. Why is that? Why is this area in the southern Central Coast of California one of the coolest wine areas in the state?

Think about this. Many of the world's wine regions either blossom along rivers, such as the Rhine River, flowing from the Swiss Alps north to the North Sea, and Bordeaux's southeast-northwest–running Gironde Estuary, or in valleys that run essentially in a north-to-south direction, such as Napa and Sonoma.

Thanks to tumultuous plate tectonic shifts over millions of years, the wine valleys of the southern Central Coast run defiantly east to west, especially Santa Barbara's three wine-growing valleys: the Santa Ynez Valley, cut east to west by the Santa Ynez River; the Santa Maria Valley, whose large agriculture plain benefits from eons of deposits from the Santa Maria River; and the vineyard-plush Los Alamos Valley, fed by many small streams.

These transverse mountains make highly efficient conduits for fog, at night and early morning, and cold offshore breezes that funnel inland from the Pacific Ocean. Thus, the cool climate extends the growing season and hang time for the grapes, which in turn gives the vintners the rare ability to harvest grapes at full bright ripeness and with extraordinary amounts of acidity.

"It's a grand slam," comments Wes Hagen. "We've got fruit blessing from California, sun and acidity from cool climate, and complexity from high-calcium soil."

While it's possible to make Pinot and Chardonnay in warmer climates, the places where they truly thrive and deliver complexity, longevity, and character are cool regions—Burgundy, Champagne, and Alsace in France; Switzerland (mainly in German cantons); Napa's Los Carneros; Sonoma; Mendocino; Oregon; New Zealand; and Santa Barbara's North County.

Today wine is grown in all over North America, meaning Canada, the United States, and Mexico. Pinot, though, is grown almost entirely within about twenty-five miles of the Pacific Ocean. From the mouth of the Columbia River to the Santa Barbara Channel and within those twenty-five miles is where more than 95 percent of North American Pinot resides.

***

In 1970, Richard Sanford, with a degree in geography from UC Berkeley, roamed the backroads of Santa Barbara County's transverse mountain ranges. He drove with a thermometer in the car. Sure enough, the temperature dropped one degree for every mile driven west.

As a naval officer during the Vietnam War, he had become "disenchanted with the culture that sent me to the war." He came home from the Philippines by way of Kolkata and Kathmandu, Nepal, on a "spirit quest," which saw him give up his religion and turn to pacifism and the study of Taoism as a belief system. He vowed to do something different in his life, something connected to land. "That was important to me."

He recalled an incident before shipping out when he went to dinner at Bully's restaurant in La Jolla with a shipmate. The man, named unbelievably Scott Wine, ordered a bottle of Volnay.

"I didn't know much about wine, but I was touched by the velvety structure and magical mouthfeel of the Pinot Noir," Richard recalls. In thinking about that taste, after his harsh experiences in the war, he believes that "subliminally, I decided to be in agriculture, and grapes were something that gets better with age rather than [being] perishable."

He and his partner, Michael Benedict, an academic and botanist, were searching for a vineyard spot anywhere along the West Coast for a vineyard project. They looked at many places but kept coming back to the western reaches of the Santa Ynez Valley.

Luck may have played as big a role as scientific guesswork in their discovery in 1971 of a big ranch halfway between Buellton and Lompoc that hadn't been farmed since World War II. Due to an ancient landslide, it contained more complex soils than those near the gravelly riverbed of the Santa Ynez River. With their backgrounds, the two made "a credible team." So Richard found wealthy investors from his college yacht-racing days, and with the "sweat equity" the two young men were willing to invest into the project, the Sanford & Benedict Vineyard came into being.

Initially, they planted Riesling and Cab because no rootstocks for Pinot were available anywhere. In 1972 they bought Pinot cuttings from Carl Wente in Livermore Valley and "an interesting clonal selection from Paul Masson" in Saratoga. Sanford & Benedict sold their early harvests, but a 1975 barrel of Pinot tasted so good that the two men converted a barn on the property to a winery, with Richard making fermenters at a friend's hot tub factory in Santa Barbara.

They made their first Pinot Noir wine in 1976. This was the eureka moment in Santa Barbara winemaking. But that "moment" wouldn't be appreciated until the release of the now famous Sanford & Benedict Vineyard 1976 estate-grown Pinot Noir. Frank Ostini recalls how that wine rang everyone's bell.

"We all got to taste it [when released] in 1978," says Frank. "The wine made the international press as one of the few Pinot Noirs in California that was Burgundian in style, an elegant and refined beautiful Pinot Noir. About ten of us in the region decided we wanted to make Pinot Noir."

It didn't hurt that wine writer Robert Balzer published an article about the Sanford & Benedict Vineyard Pinot entitled "American Grand Cru in a Lompoc Barn."

So after seven years of hard sweaty labor, Richard found himself invited to a tasting of his ballyhooed wine at the tony California Vintage Wine Society at the California Club in downtown LA. If the white tablecloths, crystal glassware, and chandeliers didn't intimidate the thirty-year-old winemaker, the serious-faced men in fancy dress and ties sitting like a jury at the tables certainly did.

No one spoke. The wine was poured. Everyone sniffed, swirled, and sipped. Silence. Then a loud voice boomed out, "Son of a bitch!"

"Oh my God, I thought, they don't like my wine," Richard recalls decades later. "I discovered this was a term of endearment. 'Finally, a Burgundy from California!' someone else said. I took a big sigh."

***

Canadian-born architect Pierre Lafond opened a wine and cheese shop in Santa Barbara's El Paseo district, a complex of historic buildings downtown. Then a light bulb went off. In 1962 he opened the Santa Barbara Winery, bottling wine brought from more northern counties. Two years later he put a winery facility on Anacapa Street that remains there today in what is now called the "Funk Zone," near the Santa Barbara train station.

By 1965, he was producing his vintages from grapes trucked in from San Luis Obispo County to the north. His was the first commercial winery in Santa Barbara County since the 1920s. Located two blocks from the Pacific Ocean, it might be the closest winery to the sea in North America.

When Bruce McGuire joined as winemaker for the 1982 harvest, things changed. His wines started winning medals from multiple wine competitions. The wines are essentially produced off seventy acres of vines between Buellton (the town Pea Soup Anderson made famous) and Lompoc (the town W. C. Fields made famous in his radio shows and 1940 film *The Bank Dick*, calling it "Lom-pock" (as in *pocket*), although Lompocals have always pronounced it "Lom-poke" (as in *coke*).

The opening of Zaca Mesa Winery in 1978, located at the base of a hill far back from Firestone Vineyard off Foxen Canyon Road, kicked off a stellar career in winemaking for Ken Brown, its head winemaker. Under his tutelage, many emerging winemakers worked at the winery; indeed, the winery developed a reputation as a postgraduate course for such top winemakers as Jim Clendenen, Adam Tolmach, Bob Lindquist, Daniel Gehrs, Rick Longoria, and Lane Tanner.

"We called it Zeca Mesa University," recalls Bob Lindquist. "We learned a lot of what to do and what *not* to do, which is just as important."

As the region grew, its big brothers in Napa and Sonoma and at other out-of-county wineries took notice. Local growers soon found themselves growing most of their grapes for wineries not in Santa Barbara. High-quality grapes were being siphoned off to enhance and bolster other regions' products. With, by some estimates, 75 percent of the grape crop heading north, the region lacked any identity for consumers or, for that matter, itself.

Further underscoring the area's identity crisis, those wine corporations that had bought Santa Barbara fruit now became purchasers of land and developers of vineyards. In 1987 Kendall-Jackson Wines bought Cambria and one thousand acres of vineyards in the area; Robert Mondavi acquired Byron and four hundred acres.

Enter the mavericks, guys and gals ranging from Wild Man Jim, looking more like a biker than a vintner, and Frank, in his ubiquitous pith helmet, bottling small batches of Pinot and other varieties for his family restaurant, The Hitching Post, to former TV star Fess Parker in his coonskin hat and the "Pinot Czarina" Lane Tanner, a former Hitching Post winemaker and ex-wife to Frank, who established a one-woman Pinot winemaking operation and thought nothing of posing for a photo in a bubble bath with only a bottle of Pinot Noir for adornment.

Without a track record and working in a region suffering from an identity crisis, these mavericks were more open to new varieties and blends. Chardonnay was Santa Barbara's first success, becoming the area's best-selling grape. While the Sanford & Benedict Vineyard laid the groundwork for all that followed, it was Jim, a bulky man with long blond hair, a Van Dyke beard, and loud shirts, who gave Santa Barbara its much-needed identity. A one-month stay in Burgundy and Champagne in 1974 had convinced Jim to abandon his law studies and get into winemaking.

After the 1980 harvest at Zeca Mesa, Jim took off for Australia to work harvest and later that year returned to Burgundy, this time to work another, 1981 harvest. He was searching for his winemaking identity just as Santa Barbara was. While in Burgundy the renowned American wine importer Becky Wasserman hired Jim to interview her Burgundian winemakers about their viticulture and winemaking techniques.

When he returned home, he and Adam Tolmach founded their winery Au Bon Climat (ABC), or "good place" in the French winemaking idiom, in 1982 to concentrate on Burgundian-style Pinot Noir and Chardonnay. They operated out of a dairy barn for seven years before moving to a newly built Santa Maria warehouse in the foothills of the Bien Nacido Vineyards in 1989. (A year later, Tolmach left to start his Ojai Winery and Jim became the sole proprietor of ABC.) By the mid-eighties, thanks in large part to ABC, Pinot was established as the red grape in Santa Barbara County.

Jim's importance to Santa Barbara wine "can't be exaggerated," insists Wes. "Not only did he and Ken Brown at Byron put Pinot Noir on the map, but what made Jim so special was Ken was a homebody in the cellar. Jim took it to the world. He was on the road two hundred days of the year, and that meant Burgundy, Japan, China, Germany. He got on a plane and brought Santa Barbara to the world."

"He found a market internationally for Chardonnay and Pinot Noir because of his relentless pursuit," agrees Bob. "He was a road warrior carrying the Santa Barbara County flag all over the world."

"Jim befriended every fine chef in the world," says Frank. "He'd go to Finland for the best restaurant in the world. Jim was an international traveler. He promoted all of us in a wonderful way."

Jim, with ABC and later Clendenen Family Wines, put the region on wine lists and retail stores with his refined style of Pinot Noirs and Chardonnays. "Jim always said he was making a product that was a great alternative to Burgundy—age-worthy Chardonnay and Burgundian-style Pinot," notes Frank. "When Burgundy gets crazy expensive, his stuff will be so attractive, he insisted, and that's happening now."

The beating heart of ABC's funky warehouse of a winemaking facility is the open kitchen anchored by a long farm table, where Jim would tower over a large pot, simmering a stew laden with spices and good doses of red wine, whipping up his daily family-style lunches for his staff and guests. Guests ranged from former Dodgers baseball players (Bob is a huge Dodgers fan) to world-renowned sommeliers, British and American wine writers, and fellow chefs.

Wes recalls his first encounter with Jim: "As a young winemaker I met him at The Hitching Post in 1995 or 1996 and I told him I wanted to make Pinot Noir for a living. He said, 'You know there are drugs you can take to prevent that.' I said, 'I've tried them all in college and I'm still in.' 'All right then,' he said, 'we should have a conversation.'"

Another time Wes received a dismal seventy-nine score from the world's most famous wine critic, Robert Parker, for his first stainless-steel Chardonnay. "When I next saw Jim, he said he saw the seventy-nine score, gave me a hug, and said, 'You're a real winemaker now, motherfucker!' He knew that taking one on the chin from Parker gave encouragement. It was clearly not a failure. I was making a style of wine against what Robert Parker wanted."

Jim Clendenen cooking in his Au Bon Climat winery's kitchen, Santa Maria Valley, Santa Barbara County, California AUTHOR'S COLLECTION

Foxen Canyon Wine Trail, Santa Barbara County, California AUTHOR'S COLLECTION

Richard Sanford at his Alma Rosa Estate ranch, Buellton, California
AUTHOR'S COLLECTION

Jim's winemaking philosophy was all about that perfect balance between alcohol and acidity to produce wines to enhance meals. He disliked high-alcohol wines: "They don't go well with food and conviviality. We want our wines to be energizing and stimulating and not the kind where you fall asleep at the dinner table."

"His wines were very straightforward," says fellow Central Coast vintner Gary Eberle. "Jim made wines that people liked. You combine his spectacular marketing with his winemaking capability and you've got a superstar."

Yet Pinot back then was and still is an expensive beverage, because yields are low compared to other varieties. By the early years of the new century, economic forces conspired to cause problems for Santa Barbara winemakers: overplanting, a shrinking distribution chain, and what Jim called "a general malaise among wine producers."

Then a movie shook everything up and the region experienced another boom in the boom-and-bust cycle that has existed historically in the wine trade.

"Sometimes fads take over; witness Chardonnay in the late 1970s and Merlot in the 1990s," wrote Jim Clendenen in his foreword to a wine guide in 2007. "For Santa Barbara County, the popularity of . . . *Sideways* defined another reemergence. It's ironic that a wine movie can spark a wine fad that can relaunch a region. Yes, the attention turned through serendipity, perhaps of a fad, a movie, or a moment, but the area's potential, largely fulfilled, will keep the momentum going forward."

Pinot Noir from Santa Barbara County, California

The region's 270-plus wineries are spread out over a sixty-nine-mile-long and forty-five-mile-wide region, rich with over seventy different wine grape varieties. Pinot Noir tops the list. The wine trails meander through seven American Viticultural Areas (AVAs), but the Sta. Rita Hills and Santa Maria Valley AVAs are prime Pinot country.

Sta. Rita Hills

Tucked along the southwestern edge, this region's proximity to the Pacific Ocean shrouds the land with morning fog, which lifts midmorning only to be followed by an afternoon cool breeze. The maritime climate and ancient diatomaceous soil create a unique terroir, which gives the region's Pinot Noir and Chardonnay their hallmark expressions.

Our suggested Pinots of Sta. Rita Hills:

Alma Rosa El Jabali and Caracola
Babcock Ocean's Ghost
Brewer Clifton Machado, Hapgood, and 3D
Domaine de la Côte Bloom's Field
Fess Parker Winery & Vineyard Ashley's Vineyard and Pommard Clone
Fiddlehead Cellars Fiddlestix Vineyard and Seven Two Eight
Foley Estates JA Ranch
Ken Brown Rita's Crown
Kessler Haak Ohana
LaFond Winery & Vineyards Pommard Clone
Melville Anna's Block
Sandhi Romance
Sanford Winery & Vineyard Sanford & Benedict
Sea Smoke Southing
Racines Sainte Rose and Sta. Rita Hills Cuvee
Rockhound Radian Vineyard

Pinot Noir from Sta. Rita Hills, Santa Barbara County, California © ZW IMAGES

Pinot Noir from Santa Maria Valley, Santa Barbara County, California © ZW IMAGES

Whitcraft Winery Pence Ranch Clone 459

Santa Maria Valley

Santa Maria Valley, the region's first officially approved AVA, is sprawled on the northernmost edge of Santa Barbara County. Enjoying cool ocean breezes, the wind-swept region is blessed with complex soils and diverse microclimates.

Our suggested Pinots of Santa Maria Valley:

Au Bon Climat Knox Alexander and Bien Nacido Vineyard
Byron Nielson Vineyard
Cambria Julia's Vineyard
Foxen Julia's Vineyard and Block 8
The Hitching Post Highliner and Bien Nacido Vineyard
Native9 Rancho Ontiveros Vineyard

5

Alexander Payne

No Laughing Matter?

When Jack Nicholson received the Golden Globe for Best Actor in a Drama from the Hollywood Foreign Press for his performance in Alexander Payne's *About Schmidt*, he took the stage at the Beverly Hilton somewhat puzzled. "I'm a little surprised," he told the crowd. "I thought we had made a comedy."

Which raises a question: does Alexander Payne actually make comedies? What are these films?

"Every time I make a movie, there is some version of that question," says Alexander. "To me, they are all comedies. Looking at them you could call them funny dramas or serious comedies . . . anyway, I don't know what the hell they are."

After his first two sharply observed satires—*Citizen Ruth* and *Election*—in which you can pin labels on the various characters (political hypocrite, helpless addict, smarty-pants student, love-starved educator), in *About Schmidt*, *Sideways*, and then *The Descendants*, Alexander moved on to dramatic comedies (not without satirical undertones, mind you) that observe and wryly comment on the human condition. Issues such as emotional dysfunction, sexual duplicity, physical and mental anguish, regret, self-delusion, and longings to connect ramble around in all of his movies.

The fact remains that the matters that sometimes confront his stressed characters are almost too sad and painful to bear. Critics have debated whether one even can call these films comedies.

Yes, one can.

"Comedy, it seems, is never the gaiety of things," noted Walter Kerr, "it is the groan made gay."

At the time he wrote this, in his 1967 critical study *Tragedy and Comedy*, Kerr was the eminent drama critic of *The New York Times*. In studying comedy, he found himself wondering about the pain inside the best comedies, how "what is funny had better not be laughed at." He wondered about the sight of Charlie Chaplin threatened by a lion and in another movie a hungry bear or Buster Keaton sinking bravely to his death in three feet of water. One laughs, of course, but one is laughing at terror and death.

"Inside the lightest of jests there seemed to be a hard and resistant core that, even as it provoked laughter, wished desperately to discredit laughter," he mused. "*Something* inside comedy is not funny. The form refuses to define itself on its own terms, defies explanation as an independent, self-contained eruption of high spirits, simply will not be claimed as an uncomplicated good companion. It does more than acknowledge an ache; it wishes to insist bluntly, even callously, on its often over-looked secret nature."

In his movies, Alexander exploits comedy's secret nature. As Keaton, Blake Edwards, and Leo McCarey, all comedy directors, discovered much earlier, you must break through the pain barrier.

"The best comedy makes no waivers," Kerr concluded. "It is *so*. And it is harsh."

***

Alexander Payne is a throwback to the so-called Movie Brats of an earlier era—to Martin Scorsese, George Lucas, Brian DePalma, Francis Ford Coppola, Paul Schrader, and Steven Spielberg, university-trained filmmakers who deeply studied the moviemakers of bygone eras and incorporated their hard-earned wisdom, styles, techniques, and even sometimes images into their own works. Alexander has wide-ranging interests in literature, Latin American culture, foreign languages, international relationships, and journalism, but throughout his life, he has avidly studied cinema. He especially watched and rewatched the great comedic works of Chaplin, McCarey, Billy Wilder, Ernst Lubitsch, and George Stevens, but also foreign film-makers such as Dino Risi and Mario Monticelli. In discussing his films and their influences, he often refers to films out of the past that he knows so intimately.

He was born in Omaha, Nebraska. This is the city and state where Alexander has chosen to set four of his films, like Scorsese's Little Italy and Barry Levinson's Baltimore.

Three of his four grandparents were Greek immigrants. Papadopoulos was his father's father's name. His father's birth certificate in Lincoln, Nebraska, read George Papadopoulos. His grandfather changed the family name in 1915.

His grandfather launched the Virginia Café in Omaha in 1920, and his father entered the business in 1953 for the last nearly seventeen years of the restaurant's life. The establishment burned to the ground in November 1969.

As a kid, Alexander collected films, mostly 8 mm and Super 8 prints of silent films. From the camera store, he would buy three- and twelve-minute versions of Universal horror films and Abbott and Costello. When he was eleven he started to order films from Blackhawk Films in Davenport, Iowa, which at the time was curated by the great silent movie preservationist David Shepard. Over a decade later, Shepard would be his silent movie professor at UCLA.

From the age of four, Alexander would constantly watch films on television, some current, but the preference was always for vintage movies. Weekly he went to the Joselyn Art Museum in Omaha to look at 16 mm reduction prints of classic movies. So in the late sixties and early seventies, he caught the renaissance at that time of the older comedies—the Marx Brothers, W. C. Fields, Laurel and Hardy. Plus, in the seventies—he graduated high school in 1979—he went to the movies every week.

"We know it now as a new golden age of Hollywood," he says appreciatively. "Those were the movies that most made me want to be a film director."

When he was a senior in high school, he says, "My buddies could not give a shit about *Star Wars*. Our discussion was which was better, *Annie Hall* or *Manhattan*?"

The old studio system was deader than Louis B. Mayer and the hoary, moralistic Production Code had crumbled under the weight of its own irrelevance. New blood had infused what Alexander considers this new golden age.

Many of the new directors of the 1960s trained originally in television: Arthur Penn, Sam Peckinpah, Irvin Kershner, John Frankenheimer, Sidney Lumet. Many of the new directors of the 1970s studied film aesthetics, history, and production in film schools. Peter Bogdanovich had been a critic/essayist, while William Friedkin had been a documentarian.

As with the cineasts of the French New Wave, some works bore the mark of spontaneous improvisation. Woody Allen, as Alexander's buddies realized, was overflowing with comic ideas. Spielberg assiduously studied the nitty-gritty of film history, but essentially he taught himself home filmmaking and snuck onto the Universal lot to learn from the pros. Robert Altman had done a good deal of work in TV and docs, but when he somehow burrowed into the American commercial system, the maverick filmmaker turned every filmmaking genre, technique, and tradition and its archetypical characters on their heads.

***

Alexander was headed for Omaha Central High, a public school "where anybody who is anybody attends high school in Omaha." Then a tornado partially destroyed his junior high in 1975. His mother, ever vigilant about education, suggested he attend Omaha Creighton Preparatory School, a Jesuit school.

As someone from a Greek Orthodox background, he bristled at the thought of attending a Catholic all-boys school. But he went for a year, liked it, and stayed.

"I appreciated my exposure to the Jesuits," he now says. "The phrase in Jesuit education is that 'we wish to produce men for others.' I fortunately was exposed to noncrazy Catholics. There is a river of thought in Catholicism on social justice and liberalism and Vatican II and Council of Medellín in 1968 and all that. So I was exposed to service-oriented Jesuit thought. They have for four hundred years been the intellectuals of the Church."

By going to Stanford University, he could continue his extracurricular film education at the school ciné clubs (film societies) and in the greater Bay Area at the movie revival houses in San Francisco and the Pacific Film Archive in Berkeley, where he remembers seeing classic Italian and Japanese films.

At Stanford, he studied not film but rather Spanish literature and Latin American history. Much of his junior year was spent at the University of Salamanca in Spain. After a quarter back at Stanford, he got a grant from the Latin American Studies department to do research for an honors thesis paper in history in Medellín, Colombia.

"So I went down to Medellín in the summer of 1982 for what I thought was going to be three months but fell in love with Colombia, fell in love with a Colombian woman, and stayed for a year," he recalls. "I got a job teaching English, went back to the states in June of '83, and immediately went to Washington, DC, where I interned at the State Department in the Bureau of Inter-American Affairs. I wanted to see what the foreign service would be like."

He flirted with journalism, too, applying and getting into the Columbia Graduate School of Journalism. But when the time came to apply to other grad schools, he knew he wanted to go to a film school.

While he applied to five film schools, the two most prominent were in the Southern California area, USC and UCLA.

"I thought it was going to be USC because so many famous white male hotshot directors had come out of there," he admits. In the spring of his senior year, having gained acceptance to both, he went for the first time to visit both campuses.

"Cost aside, I remember at USC somebody told me, 'Yeah, you should come to USC. It's just like Hollywood.' At UCLA somebody told me, 'Why would you want to go to USC? It's just like Hollywood.'"

He chose UCLA.

"USC is more of an industry feeder school," he says. "It doesn't necessarily encourage auteurship of your work. At UCLA it's one person, one film. You're expected to write, direct, edit your own work."

Fate, as he puts it, brought him together with a fellow who would become his roommate in 1989 when he broke up with his Colombian girlfriend. His name was Jim Taylor, and he would become his writing partner for most of his films.

Jim was not attending film school at the time but working at Cannon Films, a company specializing in genre flicks and Chuck Norris actioners, funding much of their slate through huge presales at the Cannes Film Festival market. Then he got a job as an assistant to the director Ivan Passer, a refugee from the Czechoslovak New Wave who made a few significant American films.

"I couldn't afford the rent on my apartment," explains Jim of his decision to change abodes. "Alexander called me early one morning and asked, 'Are you allergic to cats?' I said, 'No,' and he said, 'Good, because I've got two cats.'"

***

In June 1990 Alexander was about to graduate from the UCLA School of Theater, Film, and Television's grad school and knew few people in the business. By the next month, *everyone* in the business knew him—and many wanted to get into business with him.

As *The Los Angeles Times* explained to its readers in its July 8 edition, "Payne, 29, is one of this year's hot prospects. With luck, he could be directing his own Hollywood movie in a year or two."

Most film school grads do not get the town's paper of record predicting nearly instantaneous success. What happened was that in the previous month *The Passion of Martin*, his forty-nine-minute student film, was shown in the Melnitz Hall theater at an industry screening arranged by UCLA film school alumni. It concerns a photographer who falls in love with a woman who admires his work. Then he becomes more and more obsessed with her. Even his parents found the film a bit grim.

"Some viewers describe it as a black comedy," wrote the *Times* reporter. "Payne says it is 'a funny tragedy.'"

So in his very first industry exposure, people were already struggling to define his films. Comedy? Tragedy? Black comedy?

The film cost him around $35,000 spread over the three years he made it. He raised money from his parents, two jobs while in school, maxing out his credit cards, and $5,000 from his grandfather's will.

The day following the screening he received forty phone calls. This newly minted film school grad held meetings with agents, producers, and studio executives. He eventually hired an attorney; signed with agent David Lonner, then at ICM; and at the end of three weeks had a writing/directing deal at Universal Pictures for $125,000 and an office on the lot.

"I was right down the hall from Scott Alexander and Larry Karaszewski, who were writing *Problem Child*, and I was writing what became *About Schmidt*," says Alexander.

When he inked his deal, the Universal executives wanted to know what he wanted to write. He told them. They listened, shrugged, and left one of "this year's hot prospects" to his own devices.

"Coming out of film school I had an idea about this guy in Omaha who retires and realizes how much time he wasted in his life, like *The Graduate* but someone at the end of his working life, not at the beginning," he says. "I was paid $125,000 and what I wrote they couldn't have been less interested in making."

He would return to the script, which he called *The Coward*, ten years later. Nevertheless, $125,000 sounds like a lot of dough for a twenty-nine-year-old fresh out of film school.

"In Hollywood, you keep half of what you make, so 15 percent goes for the agent and lawyer, 33 percent for Uncle Sam, so I lived off $65,000 for the next five years. I never changed my lifestyle from that of a grad student. I was still living in Silver Lake, still paying $700 rent until I was thirty-nine actually."

Alexander then got a video job that on the surface seemed anything but propitious but proved otherwise. Playboy Video Enterprises wanted erotic films for a series called *Inside Out* that Propaganda Films made with Playboy. Propaganda Films was a leading music video company as well as a film production entity in those days. Many notable directors launched careers there, so it was an obvious place for Playboy to turn to for relatively inexpensive videos made by relatively talented directors, writers, and crew members.

Inside Out was a series of half-hour shorts of R-rated entertainment packaged together into a feature-length video. (Logline: "Playboy does to softcore sex films what HBO's *Tales from the Crypt* did for horror.") Yet on these shorts Alexander worked for the first time with colleagues who would become part of his filmmaking team.

He and his roommate Jim not only launched their fruitful collaboration as scenarists on these films but he began a long creative partnership with production designer Jane Ann Stewart and composer Rolfe Kent on these videos.

"How else do you meet directors?" laughs Jane Stewart about working on sex films. She and Rolfe Kent were asked to do Alexander's first feature, *Citizen Ruth*.

***

Citizen Ruth began as a screenplay called *The Devil Inside*. "Still a better title," Alexander grouses, annoyed to this day about his battles with the now infamous Harvey Weinstein. The Miramax cohead pushed him to change the title (which he reluctantly did) and to change the ending (which he found a way to dodge). Of his conflicts with the manipulative film executive, who well earned the moniker of Harvey Scissorhands during the time he reigned over much of the indie film scene, Alexander says that "compared to the horror stories about working at Miramax, I had a less unpleasant experience than others."

The genesis for the screenplay was a news story in *The New York Times*. It concerned an Indian woman in Fargo, North Dakota, who at age twenty-eight had been arrested yet again for huffing spray paint outside a bus station. She had already had several children taken away from her as an unfit mother. All the kids came out damaged due to her horrible substance addiction. The judge, knowing her history, charged her with a felony—endangerment to an unborn fetus—to keep her in jail while pregnant. But he told her if she needed to "go to a doctor" before getting out, he would let her out of jail.

The women at a local abortion clinic offered to give her a pro bono abortion. Initially, she agreed to the abortion. Then the Lambs of Christ, an antiabortion group that had been arrested for protesting outside Fargo's only abortion clinic, heard of her plight. They offered her $10,000 to have her baby.

"Her brother was quoted in the newspaper saying, 'Oh, my sister will go with whoever gives her the most money,'" recalls Alexander.

Jim had circled the newspaper article and handed it to his roommate, saying, "Did you read this damn thing? It's a movie." Alexander agreed.

"I saw a black comedy," says Alexander. "Immediately, two movies came to mind. One was *Ace in the Hole*, and I wanted the film to have some of the ferocity of *Viridiana*. It doesn't; it's more jokey than Buñuel was. I was hoping the film would have some of the bite of both, though."

It is typical of Alexander for any potential film story to evoke memories of past cinema classics. It's not that such a cine-literate director wishes to copy or imitate movies out of the past, but rather he seeks guideposts for his own original work. No one watching *About Schmidt* for the first time would think of Mike Nichols's *The Graduate*. Yet Alexander saw the film as *The Graduate* in reverse, where Dustin

Hoffman's Ben Braddock is not a college grad lost over his next step in life but rather an old man realizing he never had that discussion with himself when he was young and now sees only wasted time.

Of the two films in his head when he made *The Devil Inside*, the first, *Ace in the Hole* (1951), is Billy Wilder's cynical depiction of a corrupt newspaper reporter (Kirk Douglas) turning the tragedy of a man trapped in a collapsed cliff dwelling into a media circus to further his own sorry career. The other film was the great Luis Buñuel's 1961 *Viridiana*, less surreal than many of his later masterpieces yet one that also is an investigation into human nature and the folly of trying to perform works of mercy in a world indifferent to such acts.

***

"Miramax paid for it, not much, something like $2.2 million," says Alexander of his first feature. In those days Harvey Weinstein was known for financing films or acquiring them at Sundance and other festivals, for the films themselves but even more to be in business with hot filmmakers.

If so, he certainly didn't treat his new "discovery" with much respect. For the only time in his career, Alexander had casting forced upon him in two smaller roles. Things worked out all right, but it stuck in his craw. Then came the "kerfuffle" at Sundance. Weinstein and his brother Bob saw the movie before its festival debut, of course, and Harvey immediately demanded changes.

"He wanted me to tack on a happy ending and to change the title," says Alexander. "He and [producer] Cathy Konrad said *The Devil Inside* sounds like a horror movie. Well, no, it doesn't. He threatened to pull the film from Sundance. His brother Bob came up with *Precious*, which is how it was advertised I think in the [festival] catalogue. I suggested *Citizen Ruth* at the last minute, an eleventh-hour title, and that stuck."

As for the happy ending, Alexander pointed out he had no more footage out of which to create a new ending, happy or otherwise.

"I want a title card that says she [Ruth Stoops] goes to California and is getting her life together," Harvey demanded.

"I can't do that," replied the director.

"Then we'll pull it from Sundance," Harvey thundered.

"So I thought, okay, I'll do it badly," says Alexander. "The movie ends as she runs off into the sunset after running away from the abortion protests where nobody recognizes her. Five seconds later the title card comes up. So I did it inelegantly. At the premiere, there was huge applause and then the title came up and everyone went—huh? So we were leaving the screening, Mr. Weinstein could now magnanimously turn to me and say, 'That title card—I don't think you need it.'" Alexander allows himself a smirk: "Oh gee, thanks, Harvey, you're right yet again."

***

In his first feature film, the satirical *Citizen Ruth* (1996), Alexander attacks the dicey topic of abortion head-on, although one could argue that his actual topic is fanaticism, a scourge looming over the American sociopolitical landscape even more darkly now than when the film was made.

The flash focal point for an escalating conflict between pro-life and pro-choice forces in an unnamed town that is, of course, Omaha, is Ruth Stoops (Laura Dern), an irresponsible, indigent, glue- and paint-inhaling woman recently impregnated and now in jail.

As the battle for the fetus of Ruth Stoops intensifies—Alexander and cowriter Jim gradually make it clear that Ruth herself is of little interest to either group—each side deploys its songs and rituals, its methods of picketing and protesting to gain an edge in the news cycle.

The depiction of the deceit and devilry by the antagonistic forces, each righteously convinced of its virtuous cause, marvelously spoofs the intolerance that so malignantly infects the American cultural wars that rage on and on.

Critics were astonished to come upon an American film that not only talks about abortion but talks about it in such a manner, which skewers the stringency and hypocrisies that underlie all ideological fanaticism. Maybe it is, as critic Roger Ebert suggested, "a movie with a little to offend anyone who has a strong opinion on abortion."

This is a sharp satire that deliberately eschews the kind of happy ending that would see its central figure—"heroine" seems not quite the right word here—redeemed. She remains fabulously unredeemed and irresponsible in the closing shot.

Standing in a long queue to congratulate the director after the Q&A that follows Sundance screenings was Michael London. He loved *Citizen Ruth*. But the longer he waited to shake the hot filmmaker's hand, the more he thought better of it.

"I thought, I don't want to wait and have nothing to say to him," recalls Michael. "At some time in my life, I will have a project to bring to him, which means more than shaking his hand and saying, 'I'm so and so.'"

He left the line and never met Alexander that day. When he did finally meet him, it was to discuss *Sideways*.

Alexander Payne on set with Thomas Haden Church, Paul Giamatti, and Virginia Madsen PHOTO BY RACHEL FLEISCHER

Alexander Payne on set with Thomas Haden Church and Paul Giamatti PHOTO BY EVAN ENDICOTT

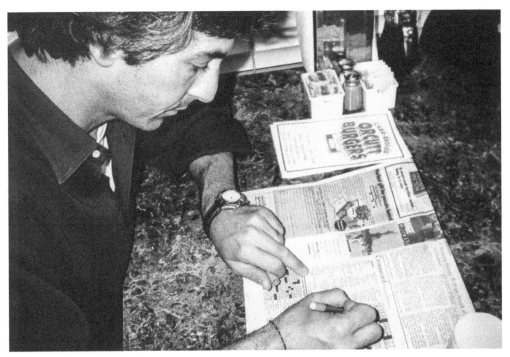

Relaxing time for Alexander Payne PHOTO BY EVAN ENDICOTT

Miramax released *Citizen Ruth* on December 13, 1996. Weinstein "dumped the movie," says Alexander. "It was not distributed worth shit. I'm not saying it would have had a big audience. But he spent zero on marketing it.

"However, even though it tanked, it garnered really good reviews. So it did what it needed to do for me: it's a good movie with a unique sense of humor and sensibility. It fit very much into that nineties New Independent Director vibe that was all the rage back then. And it scored Jim and me our next film, which was *Election*. That one really put us on the map."

***

Producers Albert Berger and Ron Yerxa brought to Alexander and Jim an unpublished manuscript called *Election* by Tom Perrotta. The project was set up at MTV Films under the aegis of Paramount.

"We thought that could make a pretty good movie," says Alexander. Within this deal, however, lurked the seeds of its box-office demise. The reason Berger and Yerxa set it up at MTV Films was because high school movies were in vogue at the time.

"The last thing in the world I wanted to make was a high school movie," says Alexander. "They thought it would fit neatly into that stream [of high school movies], but it didn't. It was neither fish nor fowl. It was an adult movie set in a high school."

Perrotta's tale of an intense high school election was inspired by the three-way American presidential campaign in 1992. A class president election finds Reese Witherspoon's overachiever Tracy Flick running unopposed. Matthew Broderick's history and civics teacher Jim McAllister, who supervises school elections, already has a wary eye on Tracy, since her seduction of a teaching colleague caused his friend to lose his job in disgrace.

So Jim persuades an affable though dim jock, Paul Metzler (Chris Klein), to run against her. Unintended consequences follow when Paul's sister Tammy decides to run as well to avenge her heartbreak when a girl she has a crush on decides to become her brother's girlfriend. Her platform is that school elections are "pathetic charades."

Jim rigs the election (and is found out) and has an adulterous affair with a neighbor—and, again, is found out. He is ostracized from his positions as a teacher and a husband. Meanwhile, Tracy, who is no less conniving and manipulative, emerges triumphant. No one has caught on to her yet—except, of course, Jim McAllister.

As with *Citizen Ruth*, all the characters are exaggerated yet also painfully real in their confusions, hypocrisies, and betrayals. Alexander refuses to take any cheap shots, though. Everyone has his or her bad moments and no one is truly the villain. Maybe Tracy will grow out of her insufferable pride. Maybe. . . .

The movie is funny, yet a viewer does wonder, should I laugh at this?

Comedy comes from physical pain as well, a situation that becomes commonplace in Alexander's film world. Ruth's intoxication is about as grim as comedy can afford. Jim sustains a nasty wasp sting on one eyelid while trying to conduct his adulterous affair. *Sideways*'s Jack endures a badly broken nose in pursuit of his philandering, not to mention an emergency early morning dash buck naked through an ostrich farm fleeing an angry husband.

In this, Alexander follows in the footpath of the late comedy auteur Blake Edwards, whose pratfalls generate sometimes excruciating pain in his dark humor.

"I would not be able to get through life had I not been able to view its painfulness in a comedic way," Edwards once said. "Leo McCarey used to talk about breaking through the pain barrier, where you're faced with so much pain it compounds itself and you can't take it anymore. So you laugh."

***

Election was a box-office failure, but for Alexander and Jim, it was a smash. Alexander explains: "We have to define 'successful.' It cost $8 million to make and officially at the box office it made $15 or 16 million. So, it's not a success financially. Yet 1999 was the first year so-called independent directors were making movies with studio money. It was considered a good year for directors in their twenties and thirties finally getting to play in a slightly different sandbox."

Those films include Kimberly Pierce's *Boys Don't Cry*, David Fincher's *Fight Club*, Sam Mendes's *American Beauty*, Paul Thomas Anderson's *Magnolia*, Spike Jonze's *Being John Malkovich*, David O. Russell's *Three Kings*, Sofia Coppola's *The Virgin Suicides*, and Steven Soderbergh's *The Limey*. Wes Anderson's *Rushmore*, which fits the pattern, had come out a year earlier.

"I'll brag," smiles Alexander. "*USA Today* ran a list of all best-reviewed movies of 1999 and *Election* was number one. Which surprised Paramount. People said the studio really didn't know what they had or how to promote it. They didn't know how to capitalize on the satiric bite of that film and also the presence of this new star Reese Witherspoon. I remember Rob Friedman [vice chair of the Paramount Motion Pictures Group], the marketing guy, said in a meeting, 'I can't market a movie called *Election*.' And he was right—he couldn't."

Election also garnered a first Oscar nomination for Alexander and Jim for Best Adapted Screenplay. This further shocked Paramount, since it came out in April, far too early for films looking for Oscars, and so it was never positioned as an Oscar film.

The Los Angeles Film Critics Association honored the two writers with its prestigious New Generation Award for that year. The duo also took Best Screenplay honors at the Independent Spirit Awards, while Alexander won Best Director and the film itself won Best Picture.

Yes, *Election* was a complete success.

6

Read This One

Hot film directors attract a mountain of reading for their next movie. But hot directors, by the very nature of being in demand and working long hours on projects, have little time to read submissions from their agents. So Alexander brought a young man in the UCLA screenwriting program on board as an intern to go through each week about 10 of his 150-odd submissions and tell him what he thought of them.

Brian Beery comes from Hollywood royalty. The Beerys are a dynasty of actors headed by the great MGM star Wallace Beery and including Noah Beery and Noah Beery Jr. Brian Beery's dad was also an actor and his mother a screenwriter, so Brian grew up in the business. He worked as a teenage actor himself in television and films.

"In the first or second week [interning for Alexander] I read *Sideways*," he recalls with still-boyish enthusiasm. "I loved it. It spoke to me. It was real and authentic."

Brian had as a theater arts/film student at UC Santa Cruz adapted and directed on the main stage an adaptation of Ernest Hemingway's 1926 novel *The Sun Also Rises*.

"In my reading, *Sideways* was close to that novel, a very deep heartfelt story of loss and longing. There's a lot of drinking, too. When I read it, I kept thinking this guy is riffing off of *The Sun Also Rises*. This [manuscript] had an authentic voice talking about a deeply personal human experience."

For his weekly report, he went to Alexander and handed him *Sideways*. "I suggest you read this one," he said.

"It went to the top of the stack," says Alexander.

In August, when he flew to the Edinburgh International Film Festival to screen *Election*, he brought along the manuscript. On the return flight, he finally picked up *Sideways*.

"It knocked me out," says Alexander. "So often I read a book and it starts well, then halfway through I'm praying, please Lord, let it continue this good. Please don't become gimmicky—don't have somebody pull out a gun. Keep it human. And this one did."

When he landed at LAX, he immediately phoned David Lonner. "I just read a future movie," he told his agent.

"The novel possessed what I look for in novels to adapt—it was lived-in," Alexander explains. "It was somehow personal to the person who wrote it. What *Sideways* has is a cri de coeur—he really needed to write that book. With Rex, he had this experience and his personal demons and he needed to exorcize them. It's very confessional, and that's what I think gives the film its idiosyncrasy and bite. The alcoholic aspect of it certainly is something that plagued Mr. Pickett, and he represented it quite faithfully but humorously in Miles. There's a real honesty about that novel that is its beauty. That book has comedy and pathos, and that combo gave Jim and me rocket fuel for that screenplay."

So Rex's story wound up in the hands of a film director, like himself, equally at home on the road. Certainly, *About Schmidt* and *Nebraska* are road movies. The *Descendants* island-hops around the Hawaiian archipelago and takes his character deeper into his life than he ever imagined. Alexander says he doesn't consciously make road movies but nevertheless jokes, "I'm the David Lean of drive-bys."

***

This is a tale of two ringing telephones. One was in Michael London's home, where he worked in a back bedroom without an assistant and without much success in his struggle to become an independent producer in Hollywood. The other was in Rex's Santa Monica house—the one with the annoying roommates.

Michael pulled into his garage on a Friday evening, returning from a discouraging meeting. Some time back he had quit a well-paying though unsatisfying job as a film executive at 20th Century Fox to see if he could get into more personal storytelling by finding creative projects through Michael London Prods. As he parked his car, he was coming to the conclusion that trying to launch a career as an indie producer was not working.

"My wife and I were running out of money," he says. "I decided time's up—go find a real job. I've got a family to support. Alexander had the [*Sideways*] novel for a while so I was ready to give up and cash in my chips. I was very disconsolate."

Hearing the phone ring, he ran into the house to catch it before the answering machine picked up. A voice on the other line said, "I'm looking for Michael London."

"This is he," replied Michael.

"My name is Alexander Payne," said the voice, "and I'd like my next movie to be *Sideways*."

Across town, Rex Pickett went for a cheap meal at Baja Fresh on Wilshire Boulevard in Santa Monica. His credit card was declined on the $6.50 charge. He trudged back to his home empty-handed. There were two messages on his ancient answering machine. The first was from his agent Brian Lifson's assistant. His recorded breathless voice demanded, "Call Brian at Endeavor ASAP." The second was from Michael London: "Just heard the news from Brian. This is extraordinary. Extraordinary!"

Now there is a minor discrepancy in the two accounts. Minor but in Rex's mind crucial to the narrative of *Sideways*. Michael says he heard from Alexander directly about his keen interest in making the *Sideways* manuscript into his next movie. In Rex's account, Michael gets the good news from the office of Brian Lifson, Rex's agent at Endeavor.

It all comes down to who submitted the manuscript to David Lonner, Alexander's agent, also at Endeavor. The call would go to whoever submitted the unpublished novel. Rex insists the submission was from Jess Taylor, his former agent, now recovering from a nervous breakdown in Brazil following his cutthroat Hollywood experiences. Michael—no one denies his wholehearted enthusiasm for the novel—believes he got the manuscript to Lonner.

What does it matter? Well, if Taylor submitted, then Michael, who never took out an option on the material, would have no paperwork as *Sideways*'s producer. Alexander could pick any producer he wanted or produce it himself. If it was from Michael London Prods. then, option or no, it came from him and he would be recognized as the de facto producer of record.

"I distinctly remember Brian [Lifson] saying, 'Wow, that was Jess's submission. You just got seriously lucky,'" says Rex.

Rex recalls getting a call soon afterward from Alexander—he even remembers his caller ID name was Constantine Alexander Papadopoulos—and after showering his novel with praise, the director abruptly asked him, "Who's Michael London and do we need him?"

When told this story many years later, Alexander laughs and says, "That sounds like something I would say. I didn't know Michael." But then he adds, after a moment's thought, "I believe it reached me with Michael London attached informally. He was serious about it, and he and I wrote checks to Rex to option it. I still have the canceled check."

David Lonner says this: "Michael London was a producer, former film executive, and journalist I had dealt with for many years. And I had come to know this was a guy with good taste, so when he talked to me about Rex's book it had credibility for me. I thought the book was great and that Alexander would be interested in it.

"What gave the manuscript credibility was Michael. He was respected by my peers and he gave me a good pitch, which got me excited about it. Rex's book delivered on the excitement of the pitch."

David doesn't remember whether or not Michael told him he had an option on the novel, but calls the whole deal a "normal transactive conversation." Many producers represent themselves as the producer of record with the author's verbal okay.

Brian, the intern who read the novel and insisted Alexander should read the manuscript, does remember the manuscript had a white Endeavor cover. So it was an interagency submission? "Yes," he says.

While the two men would go through more trials and grief before Alexander adapted the novel into a movie—*Sideways* would not be his next movie, as it turned out—the disagreement on the details of Michael's claim to the material and his eventual status as the film's producer would rupture forever whatever friendship existed between the two men. Rex feels Michael took too much credit for his discovery of *Sideways* and getting Alexander to embrace the project—"He absolutely did not get my manuscript to Lonner"—and failed to thank him later at awards ceremonies.

***

About Rex, Michael says, "When I arrived in LA he was this larger-than-life figure and I then watched him struggle to recover his artistic identity. This book is his redemption, taking all those painful experiences and turning them into something amazing."

He had read the novel in a single evening. He couldn't stop talking about it the next day, so his wife Lynn asked to read it. Michael was reluctant because the story was essentially about two middle-aged guys misbehaving around wine and women. She wanted to read it anyway.

Two hours later she was halfway through the novel, so she canceled the couple's dinner plans. She was determined to finish the book. She liked it as much as her husband. Michael now realized something.

"While the book speaks truths of men struggling with a lot of what men struggle with, it was more powerful to women because it gave women insight into male psychology and male struggles and behavior that men don't talk about much and aren't proud of. Every woman had the same reaction Lynn had—the story is funny but powerful and gives them a glimpse into the lives of men they haven't seen before.

"I thought only men would appreciate it. I think women appreciate it more."

7

Burgundy's Côte d'Or

On a crisp morning in late May 2004, the authors drive into the small village of Pommard, located south of the medieval city of Beaune, the center of the Burgundy wine trade, to visit Comte Armand Clos des Epeneaux. The winery's then twenty-nine-year-old technical director and manager, Benjamin Leroux, greets us. He is blending his 2003 vintage from two barrels—one holding wine from fifty-to-seventy-five-year-old vines and the other from thirty-to-fifty-year-old vines.

After exchanging pleasantries and reacting to our mention of the region's fabled red grape, Pinot Noir, Leroux smiles wryly. He explains he recently returned from a Pinot Noir convention in New Zealand. He admits he felt lost.

"I don't think of myself as making Pinot Noir," he tells us. "I am making Clos des Epeneaux."

Nothing more succinctly sums up the Burgundian attitude toward its famous wines than this remark. In a word, what Leroux is talking about is what the French call terroir.

The near-mystical connotation of terroir means that a wine reflects the sun and soil, the slope and climate in which the vines grow. This is fundamental to the Burgundian vigneron, but no single word exists in English for terroir. The amount of sun and rain an individual parcel or *climat* of wines receives, the pitch and composition of its earth, and the vines themselves pulling energy from the dirt below connect in a unique alchemy to create that ethereal nectar in the wineglass.

"A true Burgundy is one that expresses the terroir," says Leroux as he hands us a glass of the now-blended wine. "What I'm trying to do at Clos des Epeneaux is to express the terroir. I'm not just making Pinot Noir."

"You cannot think of the region simply in terms of Pinot Noir and Chardonnay, for in its most elemental sense, Burgundy is not about Pinot Noir and Chardonnay," writes Karen MacNeil in *The Wine Bible*. "Burgundy is about what a particular site has to say. Pinot Noir and Chardonnay are the voices through which the message is expressed."

Pinot can only express the soil of its terroir. The choice of grape was determined thousands of years ago, possibly by the Romans, because it gave the purest expression of what Burgundy meant.

Terroir and grape variety are so closely linked in Burgundy as to exclude blends of different varieties. The wines here depend almost entirely on two grapes, Pinot Noir or Chardonnay, varieties grown on five continents that trace their lineage to this region. Revered by wine aficionados, these wines can be difficult and inconsistent. A famous vineyard name on a label but from a careless grower or winemaker—a single vineyard here can have multiple owners, most notoriously Clos de Vougeot, with about eighty owners for its 125 acres—can cause bitter disappointment.

Pinot Noir is an aristocrat, yet it's a down-to-earth one because it's translating a text that is written in the soil. If you're growing the same grape throughout a region— Pinot for red and Chardonnay for white—place becomes the crucial variable. If you spend a few days tasting red wines in Burgundy, then you must realize that the grape is not totally important. Place seems to matter at least as much.

Undoubtedly, the most famous name in a region of famous names belongs to Domaine de la Romanée-Conti. The tiny, storied vineyard of Romanée-Conti produces the most expensive and, arguably, most exquisite wine in the world.

The domaine is all but hidden in the small village of Vosne-Romanée. In a narrow street, the letters RC sit atop a modest red gate, all that tip off a visitor to its august presence. Unlike wineries in Santa Barbara and elsewhere in America, the domaine is not looking for visitors.

The dapper Aubert de Villaine, proprietor of Domaine de la Romanée-Conti (popularly known as DRC), is a man who carries himself with grace and humility. Grace because he knows this earth and these vines like no one else and humility because he knows he and the legend of DRC owe everything to nature herself. Nature is his master. He is but the vineyards' caretaker.

Thus, he is a true vigneron. The French language contains no actual word for "winemaker." The French believe that man does not make wine. God does. The vigneron tends, harvests, presses, and vinifies what God—or nature if you will—has provided through the terroir. While these concepts exist throughout French wine regions, Maximillian Potter notes in his book *Shadows in the Vineyard*, "Terroir and vigneron, in Burgundy, are terms of religion, and of all the sacraments and rituals

Aubert de Villaine at the top of La Tâche vineyard, Domaine de la Romanée-Conti, Burgundy, France AUTHOR'S COLLECTION

Burgundian vignerons hold dear, none is more sacred than the marrying of a vine to the earth."

These vineyards have been in de Villaine's family since 1869, but their history dates back to about 1100, when the Benedictine monks tended the vines behind monastery walls called clos.

De Villaine greets us warmly, chats with us briefly in his office, and then drives us in his car through DRC's vineyards of Romanée-Conti and Romanée-St.-Vivant to the top of the La Tâche vineyard. The sweeping vineyards look down on Vosne-Romanée.

"People here worship the terroir because it's something that has been delivered to them—something they've inherited that is unique in the wine world and they have to defend this because it's like a work of art," says de Villaine.

As the afternoon wind kicks up at the top of La Tâche vineyard, de Villaine pays homage to the Benedictine monks who believed that vines could be cultivated in the region's limestone hillside.

"When you think that nature created this with the right exposition looking east to the rising sun and this slope creating conditions of the weather, that the monks who discovered it not only designed the vineyards but also understood the hierarchy, this heritage is so precious," he says with reverence.

Having mayoral duties in a nearby town, he drives us back to the office, where he leaves us with his cellar master who, he assures us with a twinkle in his eye, speaks English.

"*Non*," declares the cellar master defiantly. Once de Villaine departs, the man proceeds across a courtyard, down stone steps worn in the middle through centuries of footfalls to a weathered red door that leads to the barrel room. Here with a long glass tube called a "wine thief" he pulls wine samples from DRC's six-wine lineup of red wine, explaining everything to us in slow, meticulously spoken French, which we struggle to understand but somehow manage thanks to our rusty *français au lycée*.

We sample a brooding Richebourg, an exotic La Tâche, an amazing Romanée-Conti, each characterized by the cellar master as *masculine* or *féminine* or perhaps *feline* but not the sort of wine descriptions you would get in an American wine-tasting salon.

He then asks us to keep our glasses and takes us back across a courtyard to the legendary cellar, where he gives us a blind tasting to see if we were paying attention. He uncorks a dusty bottle which in typical French cellar tradition is unlabeled. (No one labels bottles in France until they are ready for sale.)

"What is this?" he demands of the wine in our glasses. At least one of us has paid attention, as she correctly identifies a feline Romanée-St.-Vivant.

"*Bon!*" He now demands for us to supply the vintage. This proves to be a bridge too far. It is the 2001 vintage, he tells us. Oh no, is he disappointed with us?

The next unlabeled and very dusty bottle he opens is more challenging, as the wine looks so light in color, like a rosé, and its fragrant, sensual aromas of rose petals fill the space around us. What is this? Could this be Romanée-Conti?

Yes, our glasses hold a 1975 Romanée-Conti, a "notorious" vintage so undistinguished at the time of release that some wine critics (mostly English) were scandalized the domaine even released it. "The worse vintage since 1968," sniffed legendary Michael Broadbent, Master of Wine and head of the wine department of Christie's auction house, when he tasted the 1975 vintage. "Poor, thin, rotten wines."

Now, nearly thirty years later, it smells and tastes like a fragile remembrance of things past, a wine that has aged gracefully into a beguiling drink that teases the taste buds and enchants the soul.

In 1980 Broadbent tasted this very wine and admitted in his notes that it had "the most attractive nose of the group [of DRC wines]—like toasted coconut; dryish, the most full bodied and far more power and intensity than the other '75s."

It turns out that tastings in this dark mystical cellar are a thing of folklore. Adam LaZarre, a veteran Paso Robles–based winemaker, gained an audience several years ago with de Villaine. In the same cellars where we stand, de Villaine brought out a bottle of unlabeled 1966 La Tâche.

"It looked like and smelled like a fino sherry," Adam recalls. "I'm thinking to myself 'Oh, this may be too old for Pinot Noir.' De Villaine looked at me, smiled, and said, 'Wait a minute.' Over the course of ten minutes, it became a ruddy light red with maybe some brown in it, then a deep sherry color, and the aroma and flavors became explosive. To the scientist in me it completely defies any logic. We open old California wines, pour a glass, and it's dead in five minutes and oxidized. This was the complete opposite. He then opened '90 Richebourg and the same thing happened. I texted a few friends of mine, winemakers in California whom I respect, who said, 'That's impossible. Wine can't do that!' I texted an importer in the UK with a Master of Wine [certificate] and he said, 'That's the magic of Romanée-Conti.'"

The village that surrounds us, Vosne-Romanée, is a small sleepy place with no cafés or stores, a post office, and a church, its streets virtually deserted, surrounded by 230 hectares of vineyards. Yet this unpretentious village—the French call it a *commune*—draws folks from around the world, wine aficionados who trek here as pilgrims to pay homage to the vineyard of Romanée-Conti. They pose in front of a centuries-old cross that marks the eastern border of the vineyards of Romanée-Conti and La Romanée (separately owned).

DRC produces seven *Grand Crus*, six of which lie in Vosne-Romanée. The other vineyard, Le Montrachet, in the Côte de Beaune, is dedicated to white wine. The entire production of these seven holdings makes up little more than sixty-two acres of vines. The total production amounts to only some eight thousand cases annually. Later in the day, we spot an older vintage of Romanée-Conti in Beaune's tony Atheneum, a wine cellar and wine bookstore, with a price tag of €3,000.

That was then. The approximate cost now of a single bottle of 2020 Romanée-Conti is $4,654.

***

There's an air of serenity in the tiny village of Volnay, whose population of four hundred is mostly made up of wine workers. Michel Lafarge, an unassuming man in his seventies with a shock of white hair, opens the door to his small winery, perched above the vineyards. Though modest in size and appearance, the winery has received international recognition for producing bottles from its twelve hectares of vineyards.

He is of a generation of vignerons who do not speak English, so through an interpreter we learn that, built over cellars first dug by the dukes of Burgundy, Domaine Lafarge was established by Michel's great-great-grandparents early in the nineteenth century. Michel Lafarge has rejected the mid-twentieth-century embrace of chemical farming, so the vineyards have never varied from traditional organic methods.

Michel, the former mayor of Volnay, leads us through the winery to one of his small vineyards in the backyard. The walled-in vineyard, Clos du Chateau des Ducs, is a *monopole* (solely owned by the Lafarge family) *Premier Cru* vineyard. To Californians used to the sweeping vineyards of Napa or Santa Barbara, it's hard to imagine that this modest "backyard" produces such internationally celebrated wines. Because the small vines are head-trained, meaning there are no wires and the plants are no more than a foot or two above the ground, one feels like one is walking in a neighbor's garden.

Later, a barrel tasting of the 2003 vintage of this *Premier Cru* reveals supple tannins. "The secret is to vinify it gently," he says. Burgundy is about purity. "The terroir should always come through—that's the skill of the vigneron."

Across from Lafarge stands the stately Domaine de Montille. We are scheduled to meet later with Étienne de Montille, a former lawyer who succumbed to the lure of the family business and now runs the domaine along with his sister Alix and father Hubert. (Hubert would pass away in 2014.)

We had run into Étienne earlier that month in Cannes, where he and his father had become unexpected stars at the world's most famous film festival after they jointly walked the famous red carpet. A last-minute selection by festival programmers, a documentary called *Mondovino*, featured several international winemakers, but the crusty Hubert and his tart-tongued quarrels with his frustrated son pretty much stole the show.

In this long-winded and preachy documentary, Jonathan Nossiter, a former Sundance Festival Grand Jury Prize winner for his black comedy *Sunday* (1997), strongly suggests that the homogenizing force of global commerce is threatening the distinctiveness of local cultures in the wine business. Folksy Hubert and his son Étienne made a vivid contrast with Napa Valley's suave Robert Mondavi, a contrast with which Nossiter hoped to prove his point about a large-scale producer versus small father-and-son farmers, although those easy caricatures were both grossly unfair and wildly inaccurate.

(In point of fact, the de Montilles, *père et fils*, were both sophisticated lawyers, while Mondavi was, unfortunately, in his dotage when his rambling and confused interview was recorded. When we asked him about the film later, Mondavi had no recollection of the filmmaker or the interview.)

Before getting reacquainted with Domaine de Montille's mouthwatering wines, however—we had already sampled some at a prescreening reception in Cannes—we get a cell phone call from Étienne.

He explains that a German friend flew in from Paris that day in his four-seater airplane. Would we like to see Burgundy from the air?

Within the hour, we are 1,500–2,000 feet in the air, peering down at villages that look like gems studded among the well-manicured vineyards. The smallness of this narrow strip of land, hemmed in by a national highway to the east—paralleled by an *autoroute* and railway line—and the trees on the upper slopes to the west, is all the more remarkable from the air, as de Montille points out the famous *communes* and domaines.

What we are looking at is a thirty-mile-long escarpment known as the Côte d'Or, arguably the most renowned wine region in the world. It starts near the city of Dijon, before which the plane banks left and slowly turns for our return trip, and extends southward to the picturesque hills around Santenay. The northern part, known as the Côte de Nuit, produces red wines almost exclusively. The southern half, the Côte de Beaune, produces both red and white wines, though whites dominate.

Etienne de Montille at Domaine de Montille, Volnay, Burgundy, France AUTHOR'S COLLECTION

Looking down, we now understand why Côte d'Or wines fetch such high prices. This narrow ridge of limestone is all there is! And where a vineyard is located on that east-facing slope nearly always gives the vineyard's rank. The less expensive *village* wines generally come from vineyards at the bottom of the slope or on the flatlands, where the soil is less well drained and mostly full of clay. Further up the stony slope or *côte* where the soil drains better, is increasingly limestone, and has the best sun exposure are the *Premier Cru* and *Grand Cru* vineyards. Over the centuries the Benedictine monks and later the Cistercians labored to identify the best soils and codify the vineyards they studied so arduously.

Cru is a French word that gets translated as "growth." A *Grand Cru* or "great growth" designation means a vineyard at the highest level of classification, while a *Premier Cru* or "first growth" isn't too shabby either.

***

Burgundy has its detractors. Wine writer Tom Maresca, writing admittedly a while ago in his 1990 *The Right Wine*, urges his reader to "bid good-bye to Burgundy. You are not losing much." He insists, "It was once a very great wine, and always costly; now it is all too frequently a very ordinary wine at a staggering price."

The brief here seems more against the *price* of Burgundy than its *taste*, but about the time he was writing, Burgundy did go into a quality slump.

He does make a valid point, though, that because "there are so many villages, appellations, hundreds of growers and *négociants*, and thousands of labels, no other group of wines presents so concentrated, and at the same time, so diverse a problem of value-for-dollar as the Burgundies."

As the legendary food writer A. J. Liebling put it, "Burgundy is a lovely thing when you can get anybody to buy it for you."

Then, too, the Burgundians almost succeeded in killing the very land on which their prized wine grew. The postwar reliance on chemical fertilizers and herbicides nearly wrecked these fine vineyards. Burgundy farmers embraced these defoliants to boost output, plus some introduced inferior high-yield clones that lead to thin, pallid, short-lived wines. Organic and biodynamic wine growing now thrives in Burgundy.

Before the early 2000s, the main challenge in Burgundy was coaxing red grapes to ripen fully. Sometimes rigorous yield management was sufficient for this purpose, as grapes ripen faster when fewer bunches are left on each vine. But in the past, more often than not, many vintages did not have enough sugar, so Burgundians would add sugar, a process known as chaptalization.

(Chaptalization is forbidden in California. However, it is permitted in Oregon, New York, and Canada. On the other hand, California winemakers may add acid, a thing that is not permitted in Burgundy.)

Then came the 2003 harvest. Six weeks of intense heat during the summer surprised growers with a shockingly early harvest. Since then, several harvests have begun in August. Thus, climate change, according to Étienne de Montille, has had a positive impact on Burgundy: "We don't need to add sugar anymore. There is enough ripeness."

The resulting wines are therefore more immediately appealing, offering fruit-forward profiles and fleshy textures coating their tannins and acid structures.

If one compares the fine (and very expensive) wines of Bordeaux to those of Burgundy, entirely different marketing strategies emerge. Bordeaux's chateaux make blends using several different varieties, such as Cabernet Sauvignon, Cabernet Franc, and Merlot, often from different vineyards. So Bordeaux in essence pitches its great wineries, such as Château Margaux or Château Latour. Burgundians focus more on terroir (a single vineyard) than on winemakers or wineries.

In Burgundy unlike Bordeaux almost all the vineyards are small. Of the *Grand Cru* vineyards owned by DRC, for example, only Romanée-St. Vivant is larger than 20 acres; Romanée-Conti itself is just 4.5 acres.

***

Why are Burgundy's vineyards so fragmented? Well, the French Revolution of 1789 ended forever the hegemony of dukes and the church as tracts of land were split up into smaller parcels and then sold off. Then came the Napoleonic Code. This gave equal rights of inheritance to all children, females as well as males, legitimate or *naturel*, in contrast to the Anglo-Saxon custom of favoring the eldest son. Thus land, particularly valuable ground such as vineyards, has been divided into smaller and smaller plots. This is something that perhaps is more desirable in theory than in practice. Remember Clos de Vougeot with its eighty owners? Napoleon is normally revered in France, but many in that country think he got this wrong.

What further complicates the search for a great Burgundy is the Burgundian system of the *négociant* or merchant. Many owners of small plots will sell their grapes to *négociants*. These *négociants* will buy grapes from different growers throughout the region. They can blend this wine with wine from other districts throughout Burgundy and bottle it under their own label. Some *négociants* have outstanding reputations, though, such as Bouchard Père et Fils, Drouhin, Latour, and Jadot.

Cellar of Domaine de la Romanée-Conti, Burgundy, France AUTHOR'S COLLECTION

The iconic cross at Domaine de la Romanée Conti, Burgundy, France AUTHOR'S COLLECTION

Route des Grands Crus, Burgundy, France AUTHOR'S COLLECTION

Burgundy's great merchant houses founded in the eighteenth and nineteenth centuries dominated the wine business for a century and a half. Then in the 1980s, some growers began to finish, bottle, and sell their own wines, offering them to importers in England and the United States. Now estate bottling and estate merchandising have become the Burgundian norm.

All wines, whether bottled by a *négociant* or a producer, will note an *appellation d'origine contrôlée* (controlled name of origin) or AOC on the label. This is a geographically based name defined by French law with stringent controls on most aspects of winemaking, not unlike the American designation of AVA or American Viticultural Area. The Côte d'Or includes eighty-four appellations.

If this sounds confusing—and it is—consider the paradoxical case of Echezeaux. As we stood on that hilltop overlooking the great vineyard of La Tâche, Aubert De Villaine pointed to a vineyard to the north. He noted that the appellation of Echezeaux, designated as *Grand Cru*, is not technically so. "The appellation had to be increased beyond Echezeaux in 1920 to please the people of Flagey [a neighboring village]. So you have some Echezeaux that sell at *Grand Cru* prices but you also have some that sell at *Premier Cru* prices."

How does one know which Echezeaux to buy?

"*Voila*, this is one of the mysteries of Burgundy that makes it difficult to understand and at the same time fascinating," admits de Villaine. "You have to search and research."

Ah, yes, search and research. So the trick for the consumer becomes not buying a bottle of Clos de Vougeot but rather knowing *whose* one or two rows of vines went into that bottle. The solution is to get to know the right person at your local wine shop, as Rex did.

Pinot Noir from the Côte d'Or, Burgundy, France

While small in size, Burgundy is one of the most intriguing and complex wine regions in the world. Its sixty-thousand-plus acres of vineyards roam through the Chablisien/Auxerrois, Côte De Nuits, Côte De Beaune, Côte Chalonnaise, and Mâconnais.

Côte de Nuits and Côte de Beaune are often grouped as the famed Côte d'Or corridor, home to such renowned names (with its wines named after the villages) as Vosne-Romanee, Nuits St. Georges, Morey St. Denis, Vougeot, and Chambolle-Musigny in the Côte De Nuits region, which some call "the Champs-Élysées of Burgundy," and Meursault, Puligny-Montrachet, Chassagne-Montrachet, Volnay, and Pommard in Côte de Beaune.

The eighty-four appellations are home to four classifications. *Grand Cru*, awarded to vineyards of highest quality wines, constitutes a mere 1 percent produced from thirty-three *Grand Cru* vineyards. For these fortunate few the name of the plot or vineyard supplants the name of the village.

About 10 percent is designated as *Premier Cru* from 640 *Premier Cru climats* or plots recognized for their high quality.

Village wines, which make up 37 percent, come from forty-four *communes* or villages. Among the better known are Mercurey, Mâcon-Villages, and Meursault.

Régional wines make up 52 percent and are produced from Burgundy's various appellations.

Also, look for *négociant* wines from Burgundy. Bottles from such outstanding houses as Bouchard Père et Fils, Maison Joseph Drouhin, Maison Louis Latour, and Maison Louis Jadot are of good quality.

Our suggested Pinots of Côte d'Or:

Château de la Crée Santenay Clos Faubard Premier Cru
Domaine Chandon de Briailles Savigny-les-Beaune Les Lavière Premier Cru
Domaine David Duband Côte des Nuits Clos de la Roche Grand Cru
Domaine Duroché Gevrey-Chambertin
Domaine Michel Lafarge Volnay
Domaine de Montille Beaune Les Perrières Premier Cru

Domaine de Montille Volnay Les Mitans Premier Cru

Domaine Thibault Liger-Belair Gevrey-Chambertin "La Croix des Champs"

Domaine de la Vougeraie Nuits-St.-Georges "Clos de Thorey" Premier Cru

Domaine de la Vougeraie Les Demodes Nuits-St.-Georges Premier Cru

Jean-Claude Boisset Vosne-Romanée Les Jacquines

Joseph Drouhin Savigny-les-Beaunes aux Clous Premier Cru

Joseph Drouhin Volnay

La Pousse d'Or Volnay Clos des 60 Ouvrées Premier Cru

Maison Evenstad, Pommard Clos de Verger Premier Cru

Simon Bize & Fils Savigny-les-Beaune Aux Vergelesses Premier Cru

Pinot Noir from Burgundy, France © ZW IMAGES

8

Sideways Gets Sidelined

If the name Harry Gittes sounds familiar, it's because that last name belongs to Jack Nicholson's protagonist in *Chinatown*, Jake Gittes. Harry Gittes, whose name did inspire the name of the detective in that classic Roman Polanski film, was an old friend of Nicholson's. He had been involved with the two movies Nicholson directed, producing *Goin' South* (1978) and coproducing and even acting in *Drive, He Said* (1971). He and fellow producer Michael Besman had found a book by Louis Begley called *About Schmidt* and sent it to his pal Jack, who expressed interest. Then a commissioned screen adaptation of the novel was presented to Alexander, now hotter than hot coming off *Election*.

"It was a serviceable adaptation of the novel which I was completely uninterested in," recalls Alexander. "That's when Jim and I said, 'Move over, Rover, let Jimi take over' [quoting a Jimi Hendrix lyric]."

Ignoring the commissioned screenplay, the writers started from scratch. Working on a story fairly similar to the one he scripted for Universal a decade earlier, *The Coward*, Alexander continued to refer back to that one so often that everyone soon agreed to blend the two. This meant Sony, the company financing the development of *About Schmidt*, needed to buy back Alexander's original screenplay from Universal.

As part of that deal, Universal got the right of first refusal on Alexander's next project. This would have ramifications when it came to making *Sideways*. Meanwhile, the two writers more or less discarded the novel and concentrated on a rewrite of *The Coward*.

Alexander explains, "We are fortunate [in our careers] in having to adapt *un*famous books so that we're able to exercise complete freedom to change at will. If you're dealing with a book that has a fan base, neither the studios nor the fans are forgiving when you take liberties."

The finished screenplay went to the two producers, Gittes and Besman. They liked it, and the following week Alexander found himself in Jack Nicholson's living room.

"I passed muster, so it was off to the races," says Alexander.

Well, not quite. The script was developed at Sony Pictures Entertainment, where it fell under the aegis of Amy Pascal, the chair of Columbia Pictures. Alexander's meeting with her did not go well. He recalls the dialogue vividly to this day:

"It's so depressing," she complained.

"But it's going to be funny," he countered.

"I want you to take me page by page and show me how it's going to be funny," she demanded.

He wasn't about to do that, so he said, 'But Jack Nicholson is going to be in it.'"

"Oh, you mean an *expensive* depressing movie," the chair retorted.

The two "trolled around" a while longer, but it was clear Sony was not going to finance the film it developed for one of the town's hottest directors and biggest stars.

Enter Rachael Horovitz, the daughter of famed playwright Israel Horovitz and sister to the Beastie Boys' Adam Horovitz. She was then a young producer and executive at New Line Cinema, a one-time independent film company that by then, since its acquisition by the Turner Broadcasting System, was regarded more or less as a major studio. She brought the project to its head, Robert Shaye. He immediately welcomed the project and saw the value to New Line of having a Jack Nicholson film. Horovitz would serve as executive producer on *About Schmidt*.

***

So now it was off to the races as a New Line film.

Again, we have another story of two telephone calls in March 2000. One evening Michael London got a call from Alexander.

"I know I told you *Sideways* would be my next movie," he began. He then told him of *About Schmidt*. He was going to make that first, then tackle *Sideways*.

Then Rex got a similar phone call. One must understand his state of mind, though, when the phone rang. "I'm on high," he remembers. "My life is going to change. Alexander is a rainmaker."

In January Artisan Entertainment made a deal to produce *Sideways* for around $10 million. Artisan was one of the more successful of the independent film and home video companies. The company hit it big with the Sundance discovery *The Blair Witch Project* (1999) and was the go-to indie around 2000. It was considered a "mini-major" studio. So *Sideways* was a greenlit picture.

His director wanted wine suggestions from Rex, as he was going to a restaurant. Then he told the novelist that he was going to make this other movie, *About Schmidt*, before *Sideways*. But he promised to do *Sideways* immediately afterward.

"I felt like somebody had hit me in the stomach," Rex recalls. "Of course, I know filmmaking and I know I'm now going to wait at least two and a half years."

He and Michael commiserated with one another.

"I remember conversations with Rex, the essence of which was—that's it. We'll never hear from him again," says Michael. "We're back where we started; the dream was extinguished. No director ever comes back to projects he was working on before he makes a first studio movie, because he's a different person and a different filmmaker when he finishes that movie."

In the immediate aftermath of that phone call, however, Rex's ex-wife won an Oscar.

Barbara Schock had gotten into the director's program at the American Film Institute (AFI). Her second-year thesis film, *My Mother Dreams the Satan's Disciples in New York*, won the 2000 Academy Award for Best Live Action Short. Her screenwriter was Rex Pickett.

"Because I've written an Oscar-winning short, I was taking meetings, getting some screenwriting work," says Rex.

Meanwhile, Alexander and Michael kept renewing their option on his novel as Alexander made his movie.

Alexander also delivered a coded message to the distraught writer and producer when he invited Michael to one of the many rough-cut screenings of his new movie. At one point in the movie, Nicholson drives his motorhome down a dusty street in a small Midwestern town (actually Nebraska City, Nebraska). On an old movie house's marquee is the advertisement "Coming Soon *Sideways*."

He's sent us a message, Michael told Rex the next day. He's not forgotten.

***

With *About Schmidt* (2002), Alexander made a significant leap in the cinematic firmament. Not only was he working with a Hollywood icon but his keenly observed insights into a troubled character and his environment, relationships, life frustrations, and regrets achieved a strong emotional impact.

Nicholson's Warren Schmidt, a recently retired actuary at a large insurance company in Omaha, suddenly finds himself widowed after forty-two years of marriage. Getting behind the wheel of a newly purchased RV he got for vacations with his wife, he hits the highway to the wedding (which he opposes) of his estranged daughter (Hope Davis) in Denver, Colorado.

Within that journey, he comes to realize the emptiness of the life he has lived so far, of his sacrifices for family and security made at the cost of his soul and humanity.

A one-sided pen-pal-ship he forms with a Tanzanian child named Ndugu, a foster child he sponsors, creates voice-overs, similar to what Alexander and Jim pulled off in *Election*, that offer insights into Schmidt's thoughts.

These are simple descriptions (confessions?) about his life and journey to a child unable to understand (or probably even read) what he is saying that give a viewer the realization that for the first time, Schmidt has come to grasp what has ailed him these many years. The anger, fear, and loneliness he suffers wash over him, making his life almost unbearable. At one moment, during the festivities of the wedding banquet, he excuses himself and moves out to the hotel's bar to order a drink so he may be alone with his thoughts.

This is a role unlike any that Nicholson, normally flamboyant with a bad-boy grin and arching eyebrows, has ever undertaken. The actor has been accused more than once of chewing scenery until there is no need to strike the set. Yet the restraint and subtlety of his Schmidt is a startling transformation. His is a failed life, and in every moment and facial expression, Nicholson wears this failure. His body feels heavy. His shoulders bow. His stoic face is made of hardened rubber.

This is a character study of a man no one would notice if passing him in a mall, a store, or an RV park. Yet in this living death of an insurance man, Arthur Miller's lines in *Death of a Salesman* ring true:

"His name was never in the paper. He's not the finest character that ever lived. But he's a human being, and a terrible thing is happening to him. So attention must be paid. He's not to be allowed to fall into his grave like an old dog. Attention, attention must be finally paid to such a person."

When Warren returns home to his accumulated mail, he discovers a "painting" from Ndugu, a child's crayon drawing of two stick figures, a man and a boy, holding hands. A tear runs down the aging man's cheek. The moment recalls E. M. Forster's opening words for *Howards End* on the title page: "Only connect . . . ". For the first time, Warren Schmidt feels that he has. Perhaps he will live his life in fragments no longer.

9

Adaptation

More often than not movies are adaptations from other media—novels, short stories, plays, comic books, manga, graphic novels, and video games, or even poetry, board games, and, in the case of the *Barbie* movie, a doll.

Alexander has made the point that he and Jim have been fortunate to adapt "*unfamous books*," so they've been able to exercise complete freedom to change the narrative and characters at will.

Yet Rex admits, "*Sideways* in somebody else's hands could have been two guys going to Cabo San Lucas doing Jell-O shots. It could have been *The Hangover*. Alexander was faithful to the locations and the characters."

Coming off *About Schmidt*, Alexander had moved into the upper echelon of directors in town. Bob Shaye had given him "final cut" on *About Schmidt*, and he would have this on all his films henceforward. Final cut means exactly that: he delivers the final edit of his film and the studio accepts this without demands for alterations. This happens only to top directors, meaning those who deliver at the box office, as Alexander had just done. *About Schmidt*'s worldwide gross was just under $106 million.

So his team—his cowriter, producer, attorney, and agent—formulated a bold plan. The script would be written "on spec." No studio paid for its adaptation.

"A director should never write his own script as a job for a studio," says Alexander. "Then they control it; you must jump through casting hoops. You've got to write on spec, then come to a studio and say, I've got this script and, ideally, these actors. You get a UPM [unit production manager] to make a budget for the film. So here's a package."

And the director gets final cut.

"This is the exact situation any representative wants," says that package's agent, David Lonner, "a highly sought-after writer-director with a great script and not tied to anyone. So we're able to set the terms. No way does a studio develop this with Paul

Giamatti and Thomas Haden Church in mind. The only reason those guys got cast in the movie is because Alexander had the freedom to put together guys he felt were those characters."

***

But first comes the great script. As the writers sat in front of a single monitor connected to two keyboards, they finally tackled *Sideways* after a long interruption.

Their methodology of adaptation is to "read the book and read the book and read the book and the moment we start writing the screenplay never pick up the book again so it comes out as cinematically as it can based on your memory of the book," says Alexander.

Sideways, however, proved different.

"*Sideways* is the one instance where Jim and I did have the book open," says Alexander. "Much of the dialogue is Rex's. We were pretty faithful to it and the story's progression. Rex has a specific sense of dialogue and humor, which we saw fit to retain."

The novel takes the form of a buddy story. Alexander, as always, thinks back to movies that move him: "A buddy film calls to mind other buddy films, *Zorba the Greek*, *Il Sorpasso*, *Withnail and I*, where at their best, particularly in *Zorba the Greek*, each [buddy] represents two sides of one human soul, a sensuous side and a more philosophical, internal side going on a trip together through life. I like that, and *Sideways* very much had that duality."

Okay, a buddy story, but what's the story really about?

It's a midlife crisis, isn't it? Drill down a notch and it's about the protagonist's search for love, for affirmation. Drill down another notch and it's about a man's passion for something, wine for sure but also validation as an artist.

Who tells the story? The novel is written in first person, and the screenwriters honored this. Which means the hero, Miles, must be in every scene. In a first-person narration in a novel, the story can only go where the hero goes and see what he sees. The movie does likewise: Miles is in every scene.

The novel's two footloose characters are both Hollywood dudes, Miles an unemployed screenwriter and Jack an actor and sometimes TV director. The screenwriters change Miles to an eighth-grade English teacher, while Jack is an ex–soap-opera actor now scrounging a living by doing voice-overs in TV commercials.

Their road trip up and down the Central Coast is reduced, so virtually the entire movie takes place in the Santa Barbara wine region with the wedding set back in LA.

In the movie, we meet the bride's family, never seen and barely mentioned in the novel. They are Armenian. As Los Angeles is believed to be home to more languages, ethnicities, and nationalities than any other city in the world, Alexander wanted to highlight this uniqueness about LA by giving his movie an "ethnic spice of some sort." As a Greek American, he probably would have gone that route, but *My Big Fat Greek Wedding* had come out in 2002. So—Armenian.

Alexander and Jim alter the novel's women considerably. The more superficial alteration modifies the tasting room pourer, whom Rex calls Tessa but they call Stephanie, mostly to flesh out a role thinly characterized in the novel as an unmarried mother and a motorcycle rider. She has a mother, too.

The Miles-Maya dynamics happen much differently on-screen. Alexander does not mind so-called unlikable characters, as witnessed in his previous films, but he does demand compelling, watchable people. In the novel Maya and Tessa take the boys off to the Cedar Spa for some hot-tubbing, with Maya opening up treasured bottles of '85 La Tâche and '90 Richebourg and seducing Miles. But it turns out that Jack paid her $1,000 to bust out her best bottles and come on to Miles sexually, making her pretty much a whore.

Instead, in the movie, a mating ritual takes place in Stephanie's rustic home over dueling speeches between these two about their love for wine, with a clear subtext that these two are falling for one another. The book does *suggest* this sequence but spoils it with its thousand-dollar indecent proposal.

Jack suffers so many rude injuries in the novel that he must make multiple trips to the Lompoc hospital's emergency room, causing a young intern to crack wise about him becoming a frequent flyer. Alexander and Jim reduce this to one hospital visit, when Stephanie slams a motorcycle helmet into the lying Lothario's nose.

In the Windmill Inn's bar, Miles in the novel meets a local crazy named Brad, who later talks the two men into going wild-boar hunting with him at night. The pimple-faced hunter ends up shooting with a rifle not at the boar but rather at Miles and Jack. This sequence feels like it belongs more in *Deliverance* than *Sideways*.

"The wild boar scene is fantastic, but it just didn't fit," says Jim.

It might seem like a minor matter, but the vehicle for a road trip is significant. The novel sends Miles and Jack off in Miles's car, a 4Runner. The screenwriters changed this to a Saab Turbo.

"The casting of a lead's car is always very important," insists Alexander. "The Ford Festiva that Matthew Broderick drives in *Election* is an example—it's the car of an impotent man. The car is key. Why a Saab? Jim and I didn't think Miles would be a 4Runner kind of guy. A Saab is something cool ten years before, that Miles still drives."

***

The screenwriters did choose to share their adaptation with the novelist, which is not their usual MO, but *Sideways* was not their usual adaptation.

"With Rex, as a courtesy, since it was so very personal, I do remember giving him the first draft after sixteen weeks," says Alexander.

Something about the porch scene bothered the novelist, however.

"Miles has his big speech about Pinot Noir," Rex recalls. "Maya says nothing. After the third draft I said, 'Alexander, she has to say something.' He didn't say anything, but in the fourth draft there was her speech."

Alexander's longtime editor Kevin Tent, who was also getting each new draft to read, noticed the change immediately: "God, it made such a difference," he recalls. It's hard to imagine Virginia Madsen getting nominated for a Best Supporting Actress Oscar without that soliloquy.

Of the wine speeches, Alexander now reflects that they "allowed me to write a lot of things personal to me about my appreciation of wine. I don't think Jim will mind my saying that I wrote more of that speech than he did, and it's a lot about my feelings about wine. Also, earlier in sequence, when in Stephanie's kitchen Miles asks Maya, 'What bottle did it for her?,' she says an '88 Sassicaia. That's the wine that did it for me. I was in Italy with an Italian girlfriend and her brother brought an '88 Sassicaia for Christmas. I never knew wine could taste that good."

***

Then there's the ending. In the novel, Maya shows up at the wedding reception in Paso Robles, holds out her arm to Miles, and says, "Come on, Miles. We don't belong here." The End.

Something about that probably felt a little too pat if not unlikely to the screen adapters. Even as they were writing their screenplay, interestingly enough, Rex was busy rewriting his novel as well.

"He had not published yet, so he was still working on his ending and we were working on our ending," says Jim. "He would send us drafts of his ending. We went a whole different direction than he did."

The screen ending is a masterful piece of writing that will later be greatly supported by acting, editing, music cues, and costuming choices that enhance the wistful and romantic final note to *Sideways* the movie.

After a return of some normalcy in Miles's chaotic life, he enters his San Diego apartment after a hard day of teaching. He has a message on his answering machine. He presses the play button and goes to look inside his refrigerator.

"Hello, Miles. It's Maya," says the recorded voice.

He freezes.

She acknowledges the letter he sent to her. Then she tells him she has finished reading his manuscript, the one that got rejected by every publisher.

"There are so many beautiful and painful things about it," Maya's voice continues. "Did you really go through all that? It must have been awfully hard. And the sister character—Jesus, what a wreck. But I have to say I was really confused by the ending. Did the father finally commit suicide, or what? It's driving me crazy."

In the screenwriting trade, there is always the difficult issue of how and where to dispense with exposition. Exposition is simply the information an audience needs to fully understand and appreciate a film. Then there is backstory. This is everything that has happened to a character before the film begins. Not all exposition is backstory. Backstory must happen to the character himself.

Alexander has a marvelous way of using backstory as bullets to be fired at exactly the right moment for maximum impact on how we feel about a character and his or her plight. In *Citizen Ruth*—for which he gives total credit to Jim Taylor—the crucial bit of backstory comes fifteen minutes before the end.

The antiabortionists found Ruth's mother and dragged her to the abortion clinic and handed her a mike, into which she pleads with her estranged daughter, "Ruth, don't do it. What if I had aborted you?" Laura Dern grabs a bullhorn and shouts back, "Well, then I wouldn't have had to suck your boyfriend's cock!"

"It makes you cringe but it's also a huge laugh and her entire backstory is suggested," says Alexander. "That's probably the best line in any movie I've done. For me, it's only important that an audience has the full mental picture of a story and its characters once the movie has ended. I hate to front-load that in any way. I do it in dribs and drabs along the way."

So in *Sideways*'s final sequence, we finally get Miles's backstory. The film cuts, as Maya's voice continues on to deliver an invitation to stop by and see her sometime in the future, to Miles's car as he drives to Buellton in the rain and mounts the stairs to Maya's apartment. One can only hope for a meeting that will allow these two in some way to resume their love. There is hope.

And along with hope we now, at the movie's conclusion, realize all that Miles has gone through in his life with his family—a wreck of a sister, a suicidal father—and add this to what we know about his struggles to secure an identity as a writer. Then we might think about how all these things contributed to his struggle with alcohol. We might also notice he is *not* carrying a bottle of wine for his get-together with Maya.

Pinot Noir from San Luis Obispo Coast, San Luis Obispo County

San Luis Obispo Coast

This AVA was awarded its status in March 2022. Popularly called the SLO Coast, this is where the coast meets the vines, vines that are influenced by soil, shaped by wind, cooled by fog, and kissed by the sun.

Vineyards here are planted in a unique seabed soil rich with fossilized shells and marine shale that gives the wines a razor-sharp minerality, the result of an ancient collision of the Pacific and Continental plates.

A group of some fifty wineries and vineyards are tucked along the eighty-mile-long stretch of land between the ocean and the Santa Lucia Mountain Range on California's Central Coast. Its wineries average only five miles from the Pacific Ocean.

Pinot Noir from San Luis Obispo County, California © ZW IMAGES

The SLO Coast AVA encompasses two adjoining AVAs, the inland Edna Valley and Arroyo Grande, as well as coastal vineyards scattered around Avila Beach, Pismo Beach, and Cambria.

Our suggested Pinots of San Luis Obispo:

Adelaida Vineyards & Winery HMR Estate Vineyard (Paso Robles AVA)
Baileyana La Entrada
Center of Effort COE
Chamisal Vineyards Morrito and Estate
Cutruzzola Vineyard Gloria
Derby Wine Estate Derbyshire Vineyard
El Lugar Rincon Vineyard
Laetitia Vineyard & Winery Les Galets
Niner Wine Estates Jesperson Ranch Reserve
Oceano Wines
Phelan Farm
Sinor-LaVallee Estate
Stephen Ross Wine Cellars Arete
Talley Vineyards Rosemary's Vineyard
Tolosa 1772, Stone Lion, and Pacific Wind
Windward Vineyard (Paso Robles AVA)

10

The Power of Yes

"**M**y whole career has been about taking a lower budget and even a much lower salary so I can cast whom I want for the movie," says Alexander. Of the final cast of *Sideways*, he says, "I could have made a decent movie with other people in it. But that's not the movie I wanted to see."

Sideways, of course, richly rewarded the director's stubborn persistence in casting the actors he felt ideally suited to those four roles rather than who was hot at the box office at that moment. Yet no film tested his resolve about casting choices more than this wine movie.

Once Alexander and Jim were satisfied with their screenplay, Alexander told his producer he wanted to audition actors immediately.

"He likes to hear many actors say the lines out loud and then start to find who should play that role by hearing the lines," says Michael.

However, this meant hiring a casting director, creating a casting office, and obtaining video equipment. With no studio backing the movie at this point, Michael was uncertain how to proceed.

"Well, you've got an office," his director pointed out.

Michael was operating out of a suite on what everyone in town called the Formosa Lot. This was a venerable lot at the corner of Santa Monica Boulevard and Formosa Avenue in Hollywood, originally established in 1918, home to United Artists, and then for many years Samuel Goldwyn.

"But we aren't in production, nor do we have any production apparatus," protested the neophyte producer.

Alexander shrugged: "Just have someone start answering the phone and say '*Sideways* Productions.'"

There was a room within the office suite that fit that purpose admirably. So Alexander and Michael paid for a casting director—Alexander always uses John Jackson—and video equipment.

"It was an incredible lesson for me," Michael now realizes. "Agents started calling and wanted their actor to come in to read. I realized you can will a movie into being by just starting to make it. We're taking matters into our own hands."

Actors wanted to be a part of an Alexander Payne movie and pressured their agents to follow through. "The script became very sought after in the actor/agent community," says Michael.

The cast Alexander came up with, after so many major actors sought these roles, certainly in 2004 lacked marquee value. These names meant little to the average moviegoer. Seeing the movie, they might then recall the face of Paul Giamatti from that Howard Stern movie *Private Parts* or have seen Sandra Oh in the HBO series *Arliss*. Older guys would certainly remember the dazzling figure and vibrant blond mane of Virginia Madsen from those teen films back in the eighties. TV sitcom watchers would recollect that guy from *Wings* and might even recall this name, Thomas Haden Church.

Yet those actors *were* these guys and gals in pursuit of Pinot Noir. They were relatable because they weren't movie stars or well-known actors.

Stars bring baggage to their films, and that's not a bad thing. Cary Grant was the best actor and the best Hollywood movie star ever. Directors—and audiences—wanted that baggage. He was an impossibly attractive man who was in on the joke: no one can be that attractive and still be the possible murderer in *Suspicion* or the short-sighted bone specialist in *Bringing Up Baby* or the cold manipulator in *Notorious*. He was handsome but could be ugly. He was charming but could be wily, if not dangerous. He was a star from the minute Mae West asked him to come up and see her sometime, and in the right role you got not only seemingly effortless nuances of character and a superb technical command of the actor's craft but you got a Movie Star. That's good baggage.

Gregory Peck brought baggage to *To Kill a Mockingbird*. His film persona was decency, integrity, reasonableness, and male self-assurance. That role needed such a movie star. In *Sideways*, though, Alexander Payne didn't want or need a movie star for any of the roles. What he wanted, though, were experienced, professional character actors to be Miles and Jack and Maya and Stephanie.

***

At the time of her casting, Sandra Oh was perhaps the least known to American audiences. Her early films were top Canadian films, and then gradually she began showing up in supporting roles in American movies and TV shows.

Sandra Oh and Thomas Haden Church PHOTO BY EVAN ENDICOTT

Downtime for Sandra Oh PHOTO BY EVAN ENDICOTT

She was born to a middle-class Korean immigrant family in Nepean, Ontario, located west of the central core of the Canadian capital of Ottawa. She declined a journalism scholarship at Carleton University to study drama at the National Theatre School in Montreal. Upon graduating in 1993, she made an immediate splash in Canadian television films. In 1994 she starred in two top TV films, *The Diary of Evelyn Lau* and a CBC biopic about Adrienne Clarkson, a Hong Kong–born Canadian journalist.

That same year she won the Genie Award (the Canadian equivalent of the Oscar) for Best Actress for *Double Happiness*, playing a young woman struggling to follow her dream of becoming an actress, which goes against the grain of expectations of her Chinese Canadian family.

She won a second Best Actress Genie for *Last Night* (1998), a Canadian dark comedy–drama written and directed by Don McKellar. In 2000, she made an American erotic drama, Michael Radford's *Dancing at the Blue Iguana*, about the lives of strippers at a so-called gentlemen's club, opposite Daryl Hannah and Jennifer Tilly.

She was beginning to appear in more American films, such as Steven Soderbergh's *Full Frontal* (2002) and supporting Diane Lane in *Under the Tuscan Sun* (2003).

Unlike anyone else, though, she didn't need to audition or badger an agent. When asked how Alexander cast her, she replies simply, "He asked me to do it."

Alexander had been thinking about offering the role of Stephanie to Sandra, his longtime girlfriend, whom he married in 2003. "It seemed to make sense to cast her and not just because we were engaged," he says. "I am very happy with her performance."

***

Of Paul Giamatti Alexander says, "I auditioned him in New York City and he was *so* good. Paul Giamatti makes even bad dialogue work. He can do anything. I wasn't very familiar with him. Meeting him I remember he had not much chance to prepare for the audition. He nailed it. He was so perfect."

Paul Edward Valentine Giamatti was born in New Haven, Connecticut, the youngest of three children. His mother, the former Toni Marilyn Smith, was a schoolteacher, as were all his grandparents. His father, Bart Giamatti, was a professor of Renaissance literature at Yale University. He went on to become the youngest president of the school in its storied history. When we interviewed Paul for this book on Zoom, towering beside him was a bookcase of many shelves jammed with enough volumes to stock a small library. So one can easily appreciate where the actor got his love for literature and reading.

In 1986 his father was appointed president of Major League Baseball's National League. He became commissioner of baseball on April 1, 1989. During his five-month tenure he permanently banned one of the game's best-known players, Pete Rose, a cocky firebrand as an All Star player and then manager, from baseball for betting on sports. Bart Giamatti died of a heart attack eight days later.

Fay Vincent, Bart Giamatti's friend and deputy, blamed Rose for Giamatti's death. Rose has made various private and public pleas for reinstatement and for eligibility to baseball's Hall of Fame (which his lifetime statistics certainly qualify him for). At a news conference at the Pete Rose Sports Bar & Grill in Las Vegas in December 2015, he spoke briefly but glowingly of Bart Giamatti, referring to him as one of "my three dads," along with his father and his Hall of Fame manager in Cincinnati, Sparky Anderson.

In the movie, Paul as Miles looks at a photo of himself and his late father in his mother's house.

Following in his father's footsteps, Paul majored in English at Yale. He toiled in regional theater in Seattle before returning to Yale's famed School of Drama, where he earned his master's degree in fine arts, with a major in drama.

New York theater beckoned in the mid-nineties, when he performed in Tom Stoppard's *Arcadia* and David Hare's *Racing Demon* and then later received fine notices for Eugene O'Neill's *The Iceman Cometh* and Bertolt Brecht's *The Resistible Rise of Arturo Ui* alongside Al Pacino.

Paul's breakout performance for film critics happened with *American Splendor*, which came out only a year before *Sideways*. Up until then, he had grabbed a bunch of showy supporting roles, such as Howard Stern's angry nemesis Kenny "Pig Vomit" Rushton in 1997's *Private Parts*, followed by *Man on the* Moon (1999) and *Planet of the Apes* (2001).

Shari Springer Berman and Robert Pulcini's *American Splendor*, a highly original mix of fiction and documentary reality, hit Sundance in 2003 like a bolt of creative lightening. *American Splendor* is based on the life of Harvey Pekar, a Cleveland filing clerk who created a comic book called *American Splendor*, which he wrote and the legendary comic artist R. Crumb illustrated.

In the film, we see the real Harvey Pekar, and then in the fictional story, we see Paul as Harvey Pekar. He looks not a bit like the writer but has uncannily adapted his mannerisms, attitudes, and cantankerousness.

American Splendor won Sundance's Grand Jury Prize. The LA Film Critics Association and National Society of Film Critics gave their Best Picture awards to the film at year's end.

When Paul auditioned for Alexander he knew nothing about the role or the movie and had never read the script, but he most definitely knew the work of its writer-director.

"I went in and read a scene and said this is nice—I got to meet Alexander Payne—left the room, and never thought about it again. It was a couple of months or so later my agent called out in LA at my hotel and said Alexander wanted to have dinner with me. He wanted to see if I wanted to play the part. The whole thing was crazy to me. 'What is the part?' I asked. She thought I had read the script. 'It's one of the leads,' she said. 'You got to be kidding me. That can't be.'"

At dinner, Paul says, Alexander asked in "his very polite Midwestern way" if he would like to play this role. He readily agreed.

"I was very skeptical, though, that he would get anybody to make the movie with me in it. He said, 'I'm interested in this guy Tom Church. You know him?' I said yeah, but I wanted to say to him, 'Good luck getting the movie made with me and him in it.'"

This charming tale about the biggest break in an actor's career omits the fruitless months Alexander spent auditioning everyone he and his casting director could think of—and being lobbied by top agents to cast major stars—in his increasingly frantic effort to find his Miles.

"When you can't find the right actor for a part, you keep auditioning, auditioning, and no one seems right, you start questioning the dialogue: maybe it's not that well written; maybe that's the problem," remembers Alexander. "Then someone walks in the door and does the version of how you heard it in your brain—the rhythm, sense of humor, and drama. Thank God."

***

Coming up with a Jack wasn't easy either. For one thing, big Hollywood stars sought the role, a role which Alexander didn't think would look good on a big Hollywood star. One of those early aspirants was George Clooney.

Alexander met with Clooney at Bob's Big Boy restaurant in Burbank. He remembers Clooney arrived on his motorcycle.

Alexander was polite but firm in his turndown. He would, of course, send Clooney his script for *The Descendants* years later, and Clooney would agree to star in the picture. That role was a much better fit, though, in Alexander's concept of casting.

"No way you can have one of the handsomest, charismatic movie stars play a down-on-his-luck TV guy. I didn't want that kind of joke in there. It doesn't make sense to me."

As he, the producer, and the casting director cycled through endless Jacks, Michael recalls that Alexander kept saying he "wanted to find a guy crazy and unhinged and funny." Finally, they decided to call in the crazy and unhinged and funny guy from the TV show *Wings*.

Thomas Haden Church was born in Yolo County, California, near Sacramento. When at age two his biological father abandoned the family, he moved with his mother to El Paso, Texas, where his maternal grandparents lived. When his mother remarried, the boy took the last name of her new husband of Nicaraguan descent, Quesada. Later, when he began to work in radio and then as an actor, he knew that name wouldn't fit a "blond, blue-eyed, pale guy," so he took two names, Haden and Church, from his grandmother's lineage.

He worked in radio while in high school and later at a college now known as the University of North Texas. He wanted to be a DJ but took a sidestep into screenwriting, which he loved, and would eventually get a degree—he admits it took seven years—in film and TV production.

His voice-over work began long before he got into acting. He sent a radio demo to an agent in Dallas, then wound up with an acting agent, who sent him to read for one of the leads in an ensemble film shooting in Missouri.

He shot for three weeks, but the film never came out. (Nor does he remember its title.) However, the film's casting director saw his work in dailies and recommended he consider going to Los Angeles to try his luck as an actor. The casting director was prescient.

Thomas arrived in LA in January 1989. Church bunked with buddies in Long Beach and immediately got a voice-over agent and then an acting agent at William Morris. By the end of the year, he was landing roles in TV, such as on *21 Jump Street*, and was hired and then fired on *China Beach*. The firing proved fortunate, as he got a *Cheers* episode instead, and because of that appearance he landed the pilot for *Wings*.

Wings, a comedy about two brothers running an airline on the New England island of Nantucket, ran for six seasons. Church was in 123 episodes from 1990 to 1995. This was followed by two seasons of *Ned & Stacey*.

He moved back to Texas permanently in 2001. He bought a cattle ranch in Hill Country and got into the beef business. A screenplay he wrote with his writing partner David Denney he wound up directing. *Rolling Kansas* (2003) was about a road trip to find a magical forest of marijuana. It got into Sundance. This put him on the festival circuit for a while and then on the verge of another directing gig at Fox when Alexander called.

On his first audition, he told the director, "I don't think I've got snowball's chance in hell of getting this part."

"You actually do," said Alexander. "I want to do this movie with actors best for the parts. I want the actor that is Miles and the actor that is Jack."

Thomas had met Alexander for the first time on *Election*. On *About Schmidt* he nearly got cast. It came down to him and Dermot Mulroney for the role of the dim son-in-law. Thomas remembers Alexander calling him on the phone and telling him, "I can't decide between you and this other guy. But I'll tell you this: if this doesn't work out, we will work together another time."

A while later—Alexander takes a long time to cast movies—Thomas was called back for a second audition. This one would become notorious in Hollywood lore.

The audition scene occurs when Jack arrives back at the Windmill Inn in the early morning after his sexual misadventures with the zaftig waitress. Miles opens the frantically banged door and Jack bursts into the room—naked. He grabs the bedclothes to cover himself, then pleads with Miles to help him retrieve his wallet with his wedding rings.

For the audition "I asked permission to take my clothes off," says Thomas. "Everybody said okay. I started to pull off my underwear and I could see a young lady, a casting assistant, was uncomfortable, so I decided to leave that on. Let's just say I was close to naked."

Michael's jaw almost hit the floor: "He was running through the offices unclothed. It was an incredible audition, the funniest, best thing I'd ever seen."

Later, when Alexander had dinner with him, Thomas asked his director about that audition. "I assumed everyone stripped down for the audition," he says.

"Nope," Alexander told him. "You were the only one."

Two weeks later, when Thomas was returning to the ranch from a Fourth of July vacation with his girlfriend, he saw he had gotten a phone call. The voice message was to this effect: "Hey, Thomas, it's Alexander Payne calling. It would be great if you could call me back. I want to check to see if you're interested in being in my movie *Sideways*."

When he called back—*immediately*—he learned the offer came with conditions.

"I want to cast you in the part, Mr. Church, but under one condition," Alexander said.

"What?" asked the concerned actor.

"You have to obey everything I say unquestioningly," the director said. Alexander remembers hearing a long pause before Thomas, of course, agreed.

"There were at least two occasions during shooting where I gave him a direction that didn't feel right to him and [he] wanted to fight with me," muses Alexander. "And then he stopped himself and said, 'I know. I promised.'"

Thomas recalls another condition.

"Remember he'd seen me almost naked," says Thomas. "He said, 'You look too fit to be the Jack I need you to be. So be less fit.'"

When reminded of this Alexander instantly responds, "Absolutely! Any movie I make, if someone works out they've got to stop working out. I can't stand how cut and ripped everybody is [in movies]. Who's going to relate to that? All the modern movie stars, I call them muscle-bound goyim named Chris and Ryan."

As Thomas jokingly puts it, "I fit the bill for Jack—a sort of has-been TV actor." Yet, ironically, it was a fluke he was available, because far from being a has-been, he nearly had a role in what became a long-running TV series.

In May his agent handed him a script for an ABC sitcom, *Less Than Perfect*. The lead actress, Sara Rue, had worked with Thomas doing a guest appearance on *Ned & Stacey*, and she wanted him for a key role. He met with all the right folks and had the inside track for the role, but his manager told him the producers wanted him to audition.

He refused.

"They know who I am. We had an amazing meeting. If they want me to test, tell them to find another actor," he told his manager.

Had he tested and gotten the part in that series, which Eric Roberts took for sixty-five episodes, he would have been unavailable for *Sideways*. He would have been in production for *Less Than Perfect* in LA.

***

While searching for their Maya, Alexander's longtime casting director John Jackson showed the director an eight-by-ten glossy of an actress off his radar. It was of Virginia Madsen, who as a young, attractive actress in the mid-eighties had starred in several teen films. She had continued working but more and more out of the limelight.

Jackson was taken by something he saw in her eyes in the headshot.

"Doesn't it look like she's been through it?" asked Jackson. He thought the look indicated some deep life experiences, perhaps happy or perhaps painful. This was what they were looking for in Maya, a divorcée who after leaving a philandering, well-heeled husband in Santa Barbara was working as a waitress while she studied winemaking.

"We met her, and her straight-ahead, plaintive delivery and approach to acting in general and to this character in particular I found just right," says Alexander. "I found her very sympathetic."

Virginia Madsen came from a Midwestern family, several of whom are involved in the arts. Her mother, Elaine Madsen, became an Emmy-winning filmmaker and author, and her brother Michael Madsen is known for his roles in Quentin Tarantino films as well as *The Natural*, *Thelma & Louise*, *Die Another Day*, and *Sin City*.

Born in Chicago, she has acted since she was nineteen and gained early recognition in her twenties in films such as *Electric Dreams* (1984), *Modern Girls* (1986), and *Fire with Fire* (1986). David Lynch cast her as Princess Irulan in his sci-fier *Dune* (1984).

She cycled through series television and was a recurring cast member on *Moonlighting* in 1989 and *Frasier* in 1999. She starred in a well-known horror flick, *Candyman* (1992), and landed a role in Francis Ford Coppola's *The Rainmaker* (1997). In other words, an early star and now a hard-working actress.

At the point she got into the room for the *Sideways* audition—a hard meeting to get, as every actress was after the role—she admits that not many people, especially in the industry, saw her work.

"This was before streaming, so your movies went straight to video," she says. "The only time people saw me was on a *Lifetime* movie. That was my bread and butter. It's nice to play moms. At least I was working, but no one really saw me."

Like Thomas, she had realistic expectations about landing such a sought-after part.

"My whole career I was second or third place [in auditions]," she says. "There was always somebody better for the money people. Even though I knew I did a good job, I usually wasn't the choice. [In this instance] I knew how right I was for this role."

It was a while after that audition that she got a call from the director. He wanted to meet for coffee, and because it was near her home she suggested the famed Chateau Marmont hotel above the Sunset Strip. Alexander startled her by offering her the role with a proviso.

"I really don't want you to wear any makeup," he said. "No glamour-puss stuff."

"So I can look just like me?" she asked incredulously.

"Yeah," he said.

"Cool, I'm so *so* ready for that," she grinned.

When she read through the screenplay, she knew her instincts about the role were right.

Thomas Haden Church and M. C. Gainey PHOTO BY EVAN ENDICOTT

"So many things in that script—the writing *sounded* like me, like they crawled inside my head. I knew where she lived, and I also had a husband that sounded like one [described] in the script, only without the show-off wine cellar."

One crucial piece of casting involved the enraged blue-collar cuckold who chases Miles down a suburban street at dawn. Alexander had an idea. He telephoned M. C. Gainey, an actor who had appeared in *Citizen Ruth*.

"M. C., I want you to be in my new movie," said Alexander.

"You got it, Dr. Payne," replied the actor. "Whatever you want."

"The only thing is I want you to be completely naked and run at the camera," said Alexander.

Gainey laughed so hard that he dropped his phone. Then he said he'd do it.

***

Then came the phone call. It came to Michael from an agent at CAA, home of many of the town's biggest stars.

In putting together the *Sideways* package, Alexander's team was laser-focused on the director's wish for the ideal conditions to make the movie: fifty-two days to shoot on location in Santa Barbara. This would cost probably about $15 to $17 million, which at that time seemed like a large amount of money for a small movie set in wine country. What drove up the cost was shooting with a union crew out of town. Were the shoot in LA, impossible with this script naturally, crew members would be at home and the production company would be off the hook for transportation and housing expenses. The issue, everyone knew, was that the nonmarquee cast Alexander wanted would not support such a budget.

So the phone call created a massive dilemma.

The CAA agent said two of the agency's biggest stars were willing to commit to *Sideways*. Their names were George Clooney and Brad Pitt. Clooney, turned down as Jack, was now looking to play Miles, and Pitt wanted to play Jack

How genuine was this offer? we must ask. Agents kick around a lot of ideas with producers and studios and nothing comes of it, but then major casting often gets done this way.

For his part, Alexander's agent David Lonner says this: "I don't know if the two had read the script, but the representative said, 'What about doing [the movie] with Clooney and Pitt?'"

Well, we do know that Clooney at least knew the script, because he had already approached Alexander about playing Jack. Now someone at his agency was proposing him for Miles with Pitt as Jack.

"It's not that Alexander didn't like them," says David. "Of course, he liked them and he later did a movie with Clooney, but he wrote those characters and Paul Giamatti and Thomas Haden Church were tailor fit for those roles."

Michael was in a sweat. Producing a movie with Clooney and Pitt would be a complete validation of his risky decision to go independent. That credit would mean a lot on anyone's résumé, whether the movie bombed or not. Nor did he have the confidence he could meet his director's expectations about financing and shooting conditions with a "no-name" cast.

"It was absolutely the turning point in the whole project," says Michael. The evening following the phone call, the two men had a frank discussion.

"What if I go with Clooney and Pitt?" Alexander asked.

"You'll get your $17 million," Michael told him.

"What if I don't go with them and go with Paul Giamatti and Thomas Haden Church?"

"You'll probably get $11 million."

"What do you think I should do?"

"The easiest path here is to cast two giant movie stars and get all the money you want. Ultimately, you have to make the decision."

Alexander called Michael the following morning: "I chose the $11 million path," he told his producer.

"I had no confidence we could get $11 million," admits Michael. "I think it was a bit of a flier at my end. At that point, though, we decided to take the package and go out with two wonderful actors attached."

However, first, there was the Artisan problem to overcome. Artisan had made a deal for the screenplay with Alexander attached to direct. When Alexander went off to make *About Schmidt*, the deal remained in place, but by the time he turned his attention back to *Sideways* Artisan was in financial distress.

David had a chat with Robert Cooper, who briefly held the post of Artisan CEO.

"It was a goodwill relationship that I had with Bob Cooper that allowed me to shop it elsewhere," says David. "I needed to get Alexander out of that predicament, and Bob was a gentleman."

***

The town loved everything about Alexander Payne's new project—except the cast. One top-level production executive who had a history with Michael phoned and leveled with him: her company was very interested in producing *Sideways* but not with that cast. Paul Giamatti was *not* a leading man, she explained.

The mandate was for Universal to get the first crack at *Sideways*. When Universal agreed to sell rights to *The Coward*, the real source material for *About Schmidt*, to New Line, the studio demanded first right of refusal for Alexander's next movie project.

Universal's specialty division, Focus Features, had been formed in 2002 by two astute executives/filmmakers, James Schamus and David Linde, out of a divisional merger of several Universal entities. These were two people who might look favorably on an ambitious indie project.

Schamus himself is a well-known screenwriter long associated with Ang Lee. He wrote or cowrote *The Ice Storm*; *Eat, Drink, Man, Woman*; and *Crouching Tiger, Hidden Dragon*. Linde is a film producer who has been involved in multiple-award-winning films and box-office champions that have grossed billions globally.

But with their corporate hats and not their filmmaker hats on, they ran the numbers. "Running the numbers" is a kind of algorithm-cum-voodoo studios swear by. You take a package—the actors and other imponderables—to see what it's "worth" in individual territories from North and South America to Europe, Asia, and beyond.

The numbers for the cast Alexander wanted came back at $12 million. Not nearly good enough for the director.

He explains, "The budget I was asking for was $15.8 million. When you're at that low a budget, each million makes a difference. If you're in the $60 million range, $3 million doesn't make that big a difference.

"I like comparing indie filmmaking or low-budget filmmaking to wine. When you go to buy wine in a store, every dollar more you pay in the four-to-twenty-five-dollar range makes a bit of difference. After twenty-five dollars you've got to know wine. You can get a brilliant thirty-dollar bottle and an atrocious two-hundred-dollar bottle. But up to twenty-five dollars, each dollar you pay makes a difference—a couple more days of shooting, a couple more months of cutting. So for that $3 million Universal *would not* fork over for this film, I had to say, 'Thanks but no thanks.'"

His first-look obligation to Universal was now satisfied.

All studios had a crack at *Sideways*, but it wound up, of course, at Fox Searchlight. That subsidiary was founded in 1994 by 20th Century Fox to focus on producing, acquiring, and distributing specialty films. (That division was one of the 21st Century Fox film companies Rupert Murdoch sold to the Walt Disney Company in December 2017, and today it's called simply Searchlight Pictures.) The new company brought in an up-and-coming film executive named Tom Rothman, who knew the indie-Sundance-specialty world, having served as president of production at the Samuel Goldwyn Company.

Starting with its first release, the Sundance pickup *The Brothers McMullen* (1995), the company enjoyed critical acclaim, but it didn't achieve a huge box-office hit until the British comedy *The Full Monty* (1997). Rothman moved on to become president of the main company, 20th Century Fox, in 1996, and by the time *Sideways* came to the company, he was chairman and CEO of Fox Filmed Entertainment.

With its art films foundering, Fox Searchlight brought aboard a new management team beginning in January 2000. Peter Rice, an affable and extremely bright young Englishman, became president. He began at Fox in the late 1980s and worked his way up the corporate ladder. During his presidency, he demonstrated an ability to recognize talent and to empower creative filmmakers in movie after movie. But he needed a signature hit.

His first move was to bring aboard as trusted lieutenants Nancy Utley, an expert marketer, and Steve Gilula, a distribution guru who cofounded the Landmark Theaters chain in 1974. This trio relied on gut instincts and not "numbers."

The company had hits with *Bend It Like Beckham*, which Steve saw at a cinema in London; the British crime drama *Sexy Beast*; and the Sundance oddity *Napoleon Dynamite*. But with *Sideways*, the trio found the project that would define all their future successes.

"This was what we wanted Searchlight to be," says Steve. "We were already a player, but in the context of 2000 and 2001, it was Miramax and Harvey (Weinstein) and then everybody else. We were kind of under the radar."

For Peter, Alexander was a filmmaker he had tracked since *Citizen Ruth*. He wanted Fox Searchlight to produce *About Schmidt*, but the budget, especially the specifics of Jack Nicholson's deal, didn't fit with Fox Searchlight's approach.

"I always felt very sad about that," says Peter.

So when he eagerly took the screenplay home, he was determined to read it that night. He opened a bottle of wine, had dinner with his family, then took up the script. About twenty-five pages in he glanced over at the wine bottle.

It was a Hitching Post Pinot Noir.

"I'm not sure why it was at my house at that moment or why I opened it," says Peter, who is to this day staggered by the kismet-like "complete and total coincidence" of that bottle appearing before him even as he was reading about Miles and Jack throwing back Pinots in the Hitching Post bar.

Nevertheless, as far as he's concerned, "it felt like fate. I just felt like *Sideways* was a movie we had to make. It was almost a perfect screenplay."

In his ensuing meeting with Fox Searchlight, Alexander was adamant the movie was not going to have movie stars, and yet he wanted $17 million.

"Give me seventeen and I promise to do everything I can to make it for less than that, but I want to know I have that to make it," Alexander promised.

"We said yes and got the movie," says Peter. "The power of yes is good."

Alexander is succinct on this point: "Our new friends at Fox Searchlight, my Greek *patron* Jim Gianopulos [a Greek American then the cochair with Rothman at Fox Filmed Entertainment] and the very progressive, forward-thinking, quietly brilliant Peter Rice, read the script, saw our package, and said, 'We agree with Universal—it doesn't make sense on paper—but we're going to make it anyway.' That's how the moguls of yesteryear made their great successes—go with your gut and not fucking numbers.

"What's Paul Giamatti worth in Spain?" Alexander asks in mimicry of the number runners in studio office suites. He laughs: "Morons!"

Alexander would be true to his word. He came in under budget—as he remembers, at $15.8 million.

"Jim and Peter never forgot that," he says. "They never had a filmmaker undercut what they gave him."

***

A final note on the casting: in the novel, neither male character is described in much detail. While in Freudian terms, one is all id and the other all superego without much of an ego to mediate between the two, they seem almost cut from the same mold, both drunks and womanizers, one who screwed up his marriage and lost it, and the other on a pathway to do the same. In the movie, of course, casting created huge differences between the short and scruffy Paul Giamatti and the rangy and impetuous Thomas Haden Church.

Film comedy teams often are opposed, both physically and mentally opposites. One is short and the other tall, one fat and the other lean. Oliver Hardy is incompetent but Stan Laurel is his proud equal, only more childlike than Hardy's ridiculous bully. Dean Martin may have played "straight man" to Jerry Lewis's clown, but when Martin removed himself, Lewis, with a few exceptions, seemed a spinning top without any guiding controller. He missed his other half.

Abbott and Costello continued this trend, and when teams multiplied we got the Three Stooges and the four Marx Brothers. On-screen, though, as with Miles and Jack, each clown is distinct.

Pinot Noir from Santa Maria Valley, Santa Barbara County, California
© ZW IMAGES

Pinot Noir from Willamette Valley, Oregon © ZW IMAGES

Pinot Noir from Sta. Rita Hills, Santa Barbara County, California
© ZW IMAGES

Pinot Noir from Sonoma Coast, Sonoma County, California © ZW IMAGES

Pinot Noir from Russian River Valley, Sonoma County, California © ZW IMAGES

Pinot Noir from San Luis Obispo County, California
© ZW IMAGES

Pinot Noir from Marlborough, New Zealand
© ZW IMAGES

Pinot Noir from Monterey County, California © ZW IMAGES

Merlot from Napa Valley, Sonoma County and Paso Robles, California © ZW IMAGES

Pinot Noir from Los Carneros, Napa Valley, California

© ZW IMAGES

Pinot Noir from Anderson Valley, Mendocino County, California © ZW IMAGES

Pinot Noir from Burgundy, France © ZW IMAGES

Pinot Noir from Burgundy, France
© ZW IMAGES

Alexander Payne and Thomas Haden Church
PHOTO BY REX PICKETT

Alexander Payne, Paul Giamatti, and Thomas Haden Church PHOTO BY REX PICKETT

Virginia Madsen at cast read through PHOTO BY REX PICKETT

Jim Taylor, Rex Pickett, and Alexander Payne

Sandra Oh, Thomas Haden Church, Paul Giamatti, Virginia Madsen

Thomas Haden Church and Sandra Oh on motorbike

Paul Giamatti and Thomas Haden Church PHOTO BY MERIE WEISMILLER WALLACE/© 2003 FOX
SEARCHLIGHT /PHOTOFEST

Paul Giamatti pouring wine from spit bucket as Thomas Haden Church looks on PHOTO BY MERIE
WEISMILLER WALLACE/© 2003 FOX SEARCHLIGHT /PHOTOFEST

11

The Frank Ostini Revolt

"The whole experience of making *Sideways* was like a hot knife through butter," recalls Alexander.

In the afterglow of a rousing and unexpected success, a director might be excused for forgetting a near calamity. Overall, the *Sideways* cast and crew reminiscences do back up his recollection of a collegial preproduction and bucolic shoot lasting nearly two months in the pastoral wine country. Yet there *was* that near calamity just before production began.

This would be his first feature taking place outside of his native Nebraska. The vision was going to be different. He began by hiring his longtime crew members, or what David Lonner calls Alexander's E Street Band, the reference being, of course, to the rock band that has been Bruce Springsteen's primary backing band since 1972. Along with casting director John Jackson, those coming on board were line producer George Parra, production designer Jane Ann Stewart, costume director Wendy Chuck, editor Kevin Tent, and composer Rolfe Kent.

However, he made one significant change to his band. After having used cinematographer James Glennon on all his previous films, he sent the script to Phedon Papamichael, then on holiday in Greece, and asked him if he would be his director of photography (DP).

Why the change?

Alexander says that still in the early stages of his career, he wanted to "mix things up a bit." But he adds, significantly, that young directors can learn much from their cinematographers.

"A DP teaches you how to make movies," he contends. "Think how many directors Gordon Willis taught how to make movies."

Willis, one of the best American cinematographers in the post-1960s era, worked extensively with Francis Ford Coppola, Woody Allen, Hal Ashby, Irvin Kershner, and

Alan J. Pakula, among others. A strong DP such as a Willis or a Papamichael can help even a director with three features under his belt achieve his vision with striking results.

Phedon had made films or videos for directors as diverse as Wim Wenders, Nick Cassavetes, Diane Keaton, and Bruce Wagner, an eclectic mix of Hollywood fare, TV miniseries, music videos, and documentaries. The two men of Greek heritage had known each other for two decades and nearly worked together on Alexander's UCLA film.

Back then, Phedon was advertising in *Drama-Logue*, a weekly magazine that contained casting and crew notices for stage and film. Because he promoted the fact he had his own camera, an Éclair NPR—"the camera with which John Cassavetes shot *Faces*," he proudly notes—Phedon inadvertently made a huge career jump.

"I didn't get jobs as a camera assistant or had to work my way through the ranks; I was a DP right from the beginning."

It was through *Drama-Logue* that he came to Alexander's attention in his search for a DP on *The Passion of Martin* at UCLA.

"I met with Alexander and interviewed to shoot, but he didn't hire me," Phedon relates with amusement. "It was one of the few jobs I didn't get. But I got another film by one of his classmates, and Alexander worked on that as a boom operator."

The two became friends and hung out with the same crowd of young filmmakers out of the AFI and UCLA film schools in the Silver Lake, Los Feliz, and Echo Park districts, and at those early LA coffeehouses, such as Onyx (conveniently located next to the Vista Theater) and Insomnia on Beverly Boulevard.

Phedon's father, also named Phedon Papamichael, came from a Greek family with roots in Constantinople, Turkey. In 1920 the family moved to Thessaloniki, Greece, where he was born two years later. He moved his young family to Los Angeles in 1966 to work as an art director for his cousin John Cassavetes on his landmark film *Faces*. He later worked with Cassavetes on *A Woman Under the Influence* (1974) and *The Killing of a Chinese Bookie* (1976).

Unlike Alexander, Phedon never attended film school but rather completed his education in fine arts in Munich, Germany, and began working as a photojournalist in New York, work that drifted into cinematography. Moving to LA at the behest of the Cassavetes family, he wound up in what amounted to a kind of film school— seven features for Roger Corman.

"They were all fifteen-day shoots, but we could do stylistically what we wanted," he recalls. "TV and feature films were very conventional then. So I had a period of doing liberating, stylistically advanced filmmaking as long as strippers were getting killed or there was full-frontal nudity. Roger was always saying, 'We need more full-frontal nudity!'

"My crew were all AFI students. My gaffer was Janusz Kaminski [who later shot *Schindler's List*, *Saving Private Ryan*, and many more films for Spielberg]. Wally Pfister [known for his work with Christopher Nolan] was the electrician on set and later my operator. And Mauro Fiore [who won an Academy Award for *Avatar*] was my dolly key grip."

While he lost touch with Alexander over the years, he followed his career. "I loved Alexander's work," says Phedon. When he read the script for *Sideways*, he immediately said yes.

To prep the movie the two not only had long talks but they screened movies. Alexander showed him several road movies, such as Hal Ashby's *The Last Detail* and Peter Bogdanovich's *Paper Moon*. By looking at movies together, Phedon feels "you discover the aesthetics of director."

One movie in particular Alexander showed him was *Il Sorpasso* (1962), directed by one of his favorite Italian directors, Dino Risi. "*Il Sorpasso* has a significant influence on *Sideways*," Alexander admits.

In that movie, a reckless and carefree rascal whisks away a timid law student for a wild two-day road trip outside of Rome in his Lancia Aurelia convertible, an impromptu expedition that opens up the younger man to aspects of life he so far has missed. The two are complete opposites, with Bruno (Vittorio Gassman) a compulsive braggart, loud, brash, and charming, while Roberto (Jean-Louis Trintignant) is painfully introverted and mild-mannered to a fault.

The ne'er-do-well Bruno, a dangerous driver—constantly indulging in *il sorpasso*, the aggressive practice of serial tailgating and overtaking other cars on the open road—is exposed as quite an impossible man, yet everyone, including his long-estranged wife (the delectable Catherine Spaak) and abandoned teenage daughter, seems to like him in a strange way—or at least tolerate his presence to a surprising degree. As for bookish Roberto, he experiences the "two best days of my life" with this pleasure-seeking man-child.

What's interesting here are the cinematic influences on the final movie. For Rex writing the novel, it was a Satyajit Ray film, *Days and Nights in the Forest*, about four city dwellers on holiday in the countryside of India. For Alexander and Phedon it was *Il Sorpasso*, an Italian comic drama about two Romans on a spontaneous joyride.

However, the major influence on the look of Alexander's films is the environment, according to both Phedon and production designer Jane Stewart.

"Alexander is affected by the actual reality of locations," says Phedon. "Location scouting is almost a dogma to him. If we scout a bar, he'll want to cast that woman who served us. All the real elements of a location inspire him."

"He does take time to make up his mind about locations," says Stewart. For the home of the zaftig waitress and her blue-collar husband, the filmmakers eventually found a house and a street in Lompoc.

"I had to find the most banal-looking place," says Jane. "Then I put a tow truck in the driveway. You've got to tell the story of your character."

Alexander likes his banality, although he prefers to give it a French pronunciation.

"He has this thing he calls the *banalité*," clarifies an amused Phedon. "It's an aesthetic where if a bad oil painting is above the bar, real kitsch, he insists on that being in the shot."

Alexander explains, "In all of my films I want vividly photographed *banalité*. As much as I'm working in the vernacular of American commercial narrative cinema, I still want to have a certain documentary approach, to have it look as real as it can. Even down to casting, which is why I cast a lot of nonprofessional actors. You need great faces. Also, Kodak film looks so beautiful you have to fight it and not light for beauty all the damn time.

"There's a mentality in American filmmaking that it must be made artificially beautiful, unblemished, to be worthy of being photographed in an American film. You have to actively fight it and to guard against that."

If walking onto a set the crew found it too pristine, everyone was obligated to mess it up. "It had to look lived-in," says Evan Endicott, Alexander's personal assistant.

This is why in the movie's first scene in Santa Barbara, when Miles and Jack hit the old Sanford tasting room, they encounter Chris Burroughs with his long hair and Stetson hat pouring wine, just as many wine tourists have down through the years.

"If someone like Chris Burroughs works in that location and is not going to freak out with cameras on, I'll go with what's real," says Alexander. "I like found objects."

Jane Stewart took a few cues from classic literature for choosing and dressing locations. For Maya's apartment, where she and Miles spend their first night together and later, at the movie's end, where he returns with the hope of reviving a friendship if not a romance, Jane and Alexander found the apartment in the city of Santa Barbara itself. Alexander liked it because of its banal look. Jane liked it because of its staircase.

"That's Miles's journey—he's got to climb up to his Juliet," she notes.

The Windmill Inn naturally provides an obvious totem of the hero's quixotic quest to establish his self-worth and perhaps find true love.

"I grew up in Holland, so I was familiar with them in my environment," she explains. "And then I was reminded of Don Quixote and his struggle with a windmill. I said, 'Let's put some material on it and pull out the Don Quixote theme. You layer a movie that way. So we climbed up with a raggedy thing—it was just ordinary white material made of canvas we aged and distressed before we put it up to draw attention to the actual windmill."

Alexander's movies feel "lived-in," Paul observes. "That's because of the people he picks to be in them and the people he picks to do them."

In searching for locations, though, producers and directors rarely give their screenplays to owners or officials, because everyone instantly becomes critics and will demand script changes if permission is to be granted for shooting at a particular facility. Usually, only those pages directly involving a location will end up in an owner's hand.

With *Sideways* the entire production ran into the Santa Barbara Vintner Association's pride and perhaps blinkered view of the wine business.

***

Frank Ostini's family has over seventy years in the restaurant business. As a young man, though, he had little interest in that business. But he briefly became a wine buyer and took a wine trip in 1978 to Europe and fell in love with the idea of making wine.

He points to the European tradition of "small restaurants serving the food they made with the wine they made. I wanted to do that. So I fell in love with the restaurant business."

To North County locals and visiting diners alike, Frank was a familiar sight at The Hitching Post kitchen's flaming grill, visible from most dining room tables. Outfitted in his chef whites and pith helmet—"It's the coolest hat over a grill"—he learned how to grill meats by simply watching his father.

For decades he had seen how hard his dad worked over the wood-fired grill. As a young man, he saw only hot, sweaty work. This was not for him—he wanted to have fun.

"Then when wine came along I realized how fun dining can be," says Frank. "That's why I told Alexander Payne you can't mess with this. It's our heart and soul."

Frank had not read Rex's novel, of course, as it was unpublished. Nor had he always enjoyed the wine-soaked antics of the author. Nevertheless, he agreed to a makeover of the dining room to suit the production in his restaurant, even as he grumbled about the days he had to close to accommodate the shoot.

Then he got hold of the script. He is vague about how it wound up in his hands. He read it. Then he called an emergency session of the Santa Barbara Vintners Association.

He told them they collectively would have to find a way if not to shut this film down to at least make life miserable for everyone associated with the production.

"Miles, who is an alcoholic, starts the movie by waking up with a hangover," says Frank. "They drink on the road. They meet up with his mother, who is intoxicated in the middle of the day."

The abuse of alcohol continues throughout the story. Frank readily admits he doesn't know how to read a screenplay. All he saw was the alcohol abuse and behavior he remembered from his experiences with Rex.

"You're reading this script and some of it's funny but you can't really tell," concedes Frank. "You have no idea how it's going to turn out and the director can take it anywhere he wants. He has final cut."

(A side note: it's interesting that in California, even this far outside of Hollywood, a winemaker-restaurateur understands the significance of final cut.)

How much trouble could Frank and the powerful Vintners Association have caused the *Sideways* production? Anyone who has seen the final movie will understand how much the loss of The Hitching Post, where Maya works and Miles and Jack make the connections that spring the entire plot, would have hurt. Okay, you find another restaurant but wait—what tasting rooms would be open to the production if the association truly boycotted the production? A scramble to find people to provide a hotel, tasting rooms, vineyards, and restaurants who were not members of the association or who were willing to buck its ardent wishes would have seriously stressed even the generous shooting schedule Fox Searchlight provided.

Sideways needed seventy-five shooting locations. Two weeks before shooting, not a single one was locked down.

Michael London urgently sought a meeting with Frank. He pleaded with him to rethink his reading of the script. He remembers his "speech" as if he delivered it yesterday:

"Frank, [the association] is wrong. This is not a bunch of Hollywood people coming up to make fun of local rednecks with drinking problems. It's a character story. It's going to be a great movie and we need you to help us. We need to shoot here and need you to believe us and trust us."

The truth of the matter is if no drinking problem existed within the Santa Barbara winemaking community, this would be the only such exception in the entire world of winemaking. Of course, there were drinking problems locally.

Yet, as Michael emphasized, this was not the subject of this comedy.

"If you read the script, you can see the depth of feeling and aspirations behind it. Was this guy often drunk? Yes. Was he drunk in a way that reflected poorly on the wine industry or Santa Barbara County? No. It was personal demons Miles was dealing with.

"I think some of the scenes when Miles is drunk are some of the most powerful in the movie—the spit bucket scene and Miles walking along highway [outside the restaurant]—grounding the movie in real lived experience and the sadness which Rex experienced during those years."

Another part of Frank's thinking about the situation perhaps was his sense that, despite Fox's involvement, this was a small-time affair. Of the filmmakers, he sniffs, "They were nobodies at the time."

He goes on: "It seemed like a pie-in-the-sky idea that the film would go anywhere. I was reluctant to be involved. This is my heart and soul. It's so important to us and we don't want it to be abused. So I want to talk to the director. I sent him a letter; I spelled it all out and he wasn't responding to it because he was starting to shoot. We were scheduled two weeks into the shoot. But we hadn't agreed yet. We hadn't signed the agreement."

Alexander Payne with crew at The Hitching Post bar PHOTO BY EVAN ENDICOTT

The Hitching Post restaurant sign AUTHOR'S COLLECTION

Frank finally got a meeting with Alexander, who told him that nothing in the film reflected badly on the wine community and that he, who had lost a brother to drug addiction, was not treating alcoholism lightly. He well understood how serious it was.

He then asked Frank to think about his films. He lays out the characters and their many foibles for audiences to make their own judgments. It was up to them to love or hate the characters, but there was no attack upon wine or the wine business.

Frank had a change of heart: "I realized we were okay in his hands." He went back to the Vintner's Association and asked them to trust the filmmakers.

"That was the turning point," says Michael. "But it almost killed us. If I've done anything valuable as a producer in my life, that was [the] most valuable."

Frank also agreed to let the crew into his restaurant.

The Hitching Post's walls were too white, so Jane needed to give it more texture. She sat with Frank's wife for several hours coming up with just the right wallpaper. That wallpaper is still there today.

The irony in all this is that The Hitching Post, more than any other establishment in the Santa Ynez area, benefited the most financially from the movie. And the vintners who came close to barring the film from their premises were a year and a half later putting up signage to trumpet the fact that *Sideways* was shot there.

Pinot Noir from Los Carneros, Napa Valley

Located at the southern and westernmost edge of Napa Valley and stretching westward into Sonoma County, the Los Carneros AVA straddles these two adjacent and famous regions. It is home to some three dozen wineries, most of them on the Napa side. The cool region of Carneros is influenced by marine air funneling in from San Pablo Bay and the Petaluma Gap to the west.

In the 1930s, Carneros Winery (now Bouchaine Vineyards) became the first post-Prohibition winery, followed by such pioneering winemakers as Louis M. Martini and André Tchelistcheff. The last three decades have drawn European families from France (Domaine Carneros and HdV), Spain (Artesa), and Sweden (Cuvaison) to the region.

Our suggested Pinots of Los Carneros:

Artesa Dijon Block Estate Vineyard
Bouchaine Vineyards Calera Clone
Cuvaison 90.1
Domaine Carneros The Famous Gate
HdV Ygnacia
Larry Hyde Barrell Selection Hyde Vineyard
Marchelle Wines Fourth Act

Pinot Noir from Los Carneros, Napa Valley, California. © ZW IMAGES

12

A Perfectly Shot Movie

One of the strongest and most memorable characters in *Sideways* is the wine country itself. It pervades nearly every frame of the film.

This may have been one of the factors that swayed Frank's change of heart. "[Alexander] knew the community," concedes Frank. He knew Alexander had been living in the area for several months, getting to understand who and what should be in the movie and which locations were right.

"I do this with every [location] movie now," Alexander notes. "I move there for many months so I get the feel of the place and get its rhythm right in the movie. With *The Descendants*, I lived in Hawaii six months before shooting."

None of his actors knew each other before getting cast. So Alexander asked them all to come to the wine country two weeks before shooting to rehearse and get used to the environment and each other, especially Paul and Thomas, who were to play the best of buddies.

When he was cast, Thomas telephoned Paul and the two spoke for about four hours, as Thomas recalls.

"He was hilarious and one of the most remarkable, well-read people I ever met in my life," says Thomas.

The swift friendship continued on location and in rehearsals.

"Before the shoot and during production they were just laughing and joking and hanging out with each other the entire time," remembers Alexander. "Sometimes it became a problem to get them to stop joking so we could shoot."

Before principal photography, the two channeled enough of their characters' mischievousness to perform at least one outrageous stunt of their own.

The actors discovered that Lompoc was also home to a federal prison, where Charles Manson resided at the time. The two decided they absolutely must visit the mastermind behind the infamous 1969 blood-soaked Los Angeles killing rampage.

"I really wanted to meet Charles Manson," says Thomas. "He was such a notorious figure when I grew up. It was always a fascination to me when I was a kid that this thing happened, that a guy who was such a cartoon character convinced other people to murder."

For Thomas and Paul, this led to their biggest regret of *Sideways*.

"They turned us away at the visitor's gate," says Thomas. "If you're not a family member, you're not allowed to see him."

That does sound like something Miles and Jack might cook up, though.

Meanwhile, Virginia and Sandra were getting deeper into wine education. Frank arranged for the two to punch down grapes at his winemaking facility. They also enjoyed one of Jim Clendenen's famous luncheons.

For Sandra those preproduction weeks were a "crash course to meet that region's wines and their boldness" as well as everyone coming together like a theater company: "We really lived the movie."

In today's moviemaking, twenty years later and after a global pandemic, cast read-throughs, if they're done at all, are on Zoom, and budgets and time have so shrunk that independent films get slammed together in two or three weeks.

"How long did we shoot?" Sandra asks.

Fifty-two days, she is reminded.

"That's *insane* now," she exclaims. "If I think about that [*Sideways* experience], what a frigging glorious time it was then that people had the time and a budget for an entire crew on location together to commune with each other. You can see the results of all that camaraderie on-screen."

"*Sideways* was such an organic, intimate style of filmmaking, not American," says cinematographer Phedon. "I imagine Hal Ashby would work like that, and I know Cassavetes worked like that with family and friends, shooting in his own house."

"The whole thing felt like a family thing," agrees Paul. "Alexander loves guys like Hal Ashby, and I think he probably makes a movie the way I like to think those guys made movies—relaxed, laid back, felt improvised but wasn't."

Even during production, weekends would involve movies screened at Alexander's rented house and other parties. "That created a work atmosphere that I absolutely believe made it on-screen," says Sandra.

***

A film set is all about hurry up and wait. It can be enormously stressful but also enormously tedious. There are many moving parts, but some parts must wait for others to move. Knowing this, Alexander Payne runs a tight ship in the loosest sort of environment. The thing everyone says about the *Sideways* production in retrospect is how "family-like" the whole enterprise was.

Phedon even told a new camera operator not to get into the business because of her experience on this particular production.

"It's not like this normally," he told her.

The DP elaborates: "Alexander remembers crew members' names from day one. He not only knows his driver's name. he knows his son plays football and for what team. 'How do you do that?' I asked him. 'I just pay attention,' he said.

"He makes an effort and is very loved by everyone on the set. Everyone feels a part of filmmaking, whether craft services or the van driver or prop master. He knows how to maximize the passion of the crew."

Perhaps it's reflective of his Midwestern decorum and modesty that he shares his film work—and its triumphs—with all his collaborators, even his driver and crew members.

"Everyone feels confident and everyone wants to do good for him because he acknowledges everyone's work," remarked Virginia. "One day he stopped everything and said, 'I just want to acknowledge how really good the craft services were today. The sandwiches were excellent.' We all applauded. The craft people were, like, nobody ever does that!

"He did for the background actors, too. He didn't call them extras. He said, 'I want to thank all of you because you're creating this world that we're in. Thank you for your focus and the characters you're playing.' Wow, they all thought, we're in it for [the] rest of [the] day now!"

Yet there is also a tight ship. He insists that his actors do not ad-lib. He and Jim spent months writing the screenplay, and it didn't require any rewriting by his players.

"Please recite the dialogue exactly as written," he announced right from the start. As he sees it, it's the actor's job to make the dialogue feel spontaneous, not for it to be spontaneous.

Once Virginia asked if she might say something different than what was written. He thought about it for a few moments.

"I like that idea but I think today we'll just do it as written," he decided.

Phedon recollects a moment when Thomas asked to change a line. As he recalls, it was a single word he wanted to change.

"Alexander thought about it for over a minute and said, 'Let's stick to the screenplay,'" according to Phedon. "He thought about it. Yet it was something he wrote for a reason. There was a reason he chose that word. His dialogue doesn't *feel* written, but it's very precisely chosen."

The exception Alexander did allow to his no-ad-libbing decree was, however, Thomas. He admits that at times the actor was "so brilliant in his humorous ejaculations that I could not *not* take them."

To illustrate, he recollects that at breakfast Jack says, "I get chicks looking at me all the time. All ages. Dudes too."

Alexander can't help a chuckle. "Thomas added, 'Dudes too.'"

During the same meal, Thomas says to Paul he aims to "get your bone smooched."

"That's his I believe. It's so funny—or maybe it came out in rehearsal and I incorporated it into the screenplay."

"Alexander is a benevolent dictator," says Paul. "He can get what he wants out of you without your feeling he's doing that at all."

***

If you walk onto just about any film set today and watch a take, you'll have a hard time finding the director. He's barricaded within what everyone refers to as the "video village."

Once upon a time, a director stood next to or near the camera watching his actors intently. Only when he saw the dailies the following day, though, would he see the shots as recorded by the camera. Then in the 1980s when videotape became more prevalent, a video signal from the camera could be fed to monitors so the director could see the actual shot in playback immediately afterward. The video monitor can also give the director an instant view of what the camera is seeing as the scene is taking place. Thus, a director can direct a scene, if he chooses, directly from the video village, often beneath a tent.

"Suddenly, the director is not there with you to watch tiny things and see what I am doing," says Virginia.

So she was grateful that Alexander is a throwback to days of old. He stands next to his cameraman and watches his actors as intently as Alfred Hitchcock or George Stevens once did. He sees what his actors are doing and is not far off looking at a monitor.

"He would stand next to the camera I'm operating," says Phedon. "A movie, no matter what size, ultimately really happens among the operator, director, actors, focus puller, and boom operator. Everyone else on the set is support to that. Where filmmaking happens is really a very tight, intimate group making the shots. Alexander is right next to me or sitting on an apple box looking at the actors and their performance, giving face-to-face feedback, eye contact with them, all those things so important but lost by today's technology with people being inside tents."

"In a scene between Paul and me, Alexander is so close, his presence so intimate, that you have confidence with what you're doing with the character," says Thomas. "I say he is really the third character in the scene."

"He is still one of the only directors I've ever worked with who didn't have a video village," remarks Paul. "That makes a huge difference—just one workspace and not two. So often the video village is where the really important people are doing all the really important work and everyone else is just schlepping around. [With Alexander] everybody is in the same place and doing the same job."

Says Alexander, "You've got to be right there with actors. Between 'Action' and 'Cut' I mentally wish away all the apparatus of filmmaking and think if no film crew were here—I'm just eavesdropping on this scene—would I believe it? That's my rule of thumb, my MO. If I believe it, that's good."

***

It took a while for the director and his new cinematographer to get on the same page, however.

"He's often a bit contradictory," says Phedon. "He said he wanted this to look beautiful like [Vittorio] Storaro shot it. He has these ideas but then his sense of realism and nonartificial cinematic beauty so dominate the process that it's not at all like he said. I realized that it can't all be beautiful.

"I struggled to find a language for him. It took me maybe the first two weeks of the shoot to get into a groove and see how he sees things—what's important to him."

"On *Sideways* I had the idea of making a sixties Italian comedy that kind of looked like an American film from the seventies," says Alexander. "Phedon suggested using a filter called Pearl Mist that gives a certain softness and allows some of the brighter lights to blow out and made it pretty."

"Since he wanted it to feel like a seventies movie, we shot with higher-speed film to have a little more grain," says Phedon.

Yet beauty kept interfering.

"On one shot we were setting up I said, 'It looks too pretty,'" recalls Alexander. "And Phedon said, 'No, it's a side of yourself you've been repressing. Just enjoy it.'"

Many directors come onto sets with long and complicated shot lists or even storyboard scenes to show the progression of the day's shots. Alexander lets his actors, locations, and the story tell him where the camera belongs.

"When you write something, you see it in a certain way in your head," says Alexander. "Then the locations show you how they wish to be shot. I like the eccentricities of a location. I wouldn't know how to design a set.

"[Director Michelangelo] Antonioni used to say, 'I come on set, block the scene with actors, and then make a well-shot documentary about it.' That for me is a very liberating way to think about it.

"After I block a scene with actors, we run it for the cinematographer and we exchange looks. Maybe I'll be one side of the action and he another and he'll go over here."

Phedon elaborates: "When we would rehearse, he'd stand in one corner of the room and then after rehearsal, he'd say maybe we should shoot there. Whereas I was moving around and thinking of different angles. So I would offer him other shots.

"We make decisions after we watch the actors work on the set. We watch it. We let them go. It's not hard since we're not doing great complicated shots with camera movement. So we create coverage by moving the camera a bit so someone walks into a close-up.

"He doesn't overcover things. We don't keep hacking away doing fifteen takes. Often we do maybe three takes. Typically, after the first take, he'll say, 'That was great. I think we should move on. Well, let's do one more for fun.' It gives actors a sense that, hey, we've got this and they can now be freer about this."

The day's shooting would never drag on; Alexander and Phedon were organized enough that things ended promptly. Then it was time to pull out a corkscrew.

Paul recalls a poignant evening scene where he comes stumbling out of The Hitching Post, alone and drunk. It was surprisingly hard to do, and for once he found himself doing a scene over and over. Finally, they got a last take.

"Alexander called, 'Cut.' I looked around and everybody was standing with a glass of wine. The shooting day is over. Just great. When he's done, he's done. No nineteen-hour days."

***

Virginia didn't realize the movie they were making until she watched the scene where Jack drops the bombshell to Miles that his wife has remarried. Miles grabs a wine bottle from the Saab's back seat, takes a swig, and then bolts down a hill pursued by Jack. All the while Miles is taking more huge swigs.

Watching the physical comedy as the two gamboled down the hill made her laugh.

"I thought, 'Oh, *this* is the movie we're making.' I didn't read the script as funny. I read it as more dramatic with bits of comedy like *About Schmidt*. I guess I thought it

Paul Giamatti and Virginia Madsen between takes on the set. PHOTO BY EVAN ENDICOTT

Paul Giamatti and Thomas Haden Church chat on the set PHOTO BY EVAN ENDICOTT

Thomas Haden Church on set after "beating" by Sandra Oh PHOTO BY RACHEL FLEISCHER

Paul Giamatti and Thomas Haden Church between takes on the set PHOTO BY EVAN ENDICOTT

was heavier and sadder because of Paul's character and the nature of Thomas's character. But when I saw them doing that, I thought this is fucking great!"

Whether through the happenstance of scheduling or the director's shrewdness, the first scene to feature all four actors was the picnic scene. It takes place on a hillside with the camera at a distance.

"The prop guys gave us bags of groceries," recalls Virginia. "So they set a blanket down and we were to have a picnic. As actors we wanted to know, 'Well, should we be in conflict, or what do talk about?' we asked. 'Go have a picnic!' called out Alexander. Okay! Such freedom. The sun was going down and we had only had fifteen to twenty minutes because the light begins to change. The light gets gold and [the] hills silvery and it's very romantic. We were going through all the food and inside was a bottle of wine and a corkscrew. That's a no-no as actors.

"We're looking off into the distance and call out, 'Should we open it?' 'Open the wine and drink it!' Alexander shouted. So we had a little toast and the sun was down almost immediately and we were in love. I haven't had an experience like that since. It was beautiful.

"The first gathering as a foursome—it made all the rest of the shoot easy. Very relaxed. Not a lot of sets or roles I've played did I feel that relaxed and therefore that creative. We weren't playing characters; we were these four new friends. He set the stage for us to have that relationship and be relaxed as performers."

"*Sideways* is probably one of best filmmaking experiences I've ever had," echoes Sandra Oh. "Because it's very rare that all the ingredients you need to make good cake come together—the script, direction, cast, crew, location, and the fact it was made during a time when independent studio films had budgets and support behind them.

"The people who made *Sideways* all loved film, and that spread into the love of wine as well. The explosion of what wine meant to a larger audience no one expected at all."

The drinking of real wine, as Virginia rightly says, is a huge no-no on movie sets but gradually became a necessity because of Alexander's quest for authenticity.

"I was very strict with the property master that all the colors of fake wine must look like the colors of real wine," says Alexander. "I can't stand it in movies where all the wine looks like cherry juice or Hi-C. What passes for wine in movies is usually dreadful. I needed Syrah to look like Syrah and Pinot like Pinot. What he and his department concocted with cranberry and grape juice and coloring made everyone sick. I said the heck with it, just drink real wine, so we pretty much did."

***

Los Olivos is an unusual town. A quarter century before *Sideways*, this unincorporated community within the Santa Ynez Valley was a remnant of the Old West, a tiny affair where longtime residents resided in comfortable homes, some ranch-style and a few dating from the Victorian era, but only a few shops. A Union Oil 76 filling station sat at the crossroads. There were no tasting rooms and limited eating establishments. Then wine tourism claimed the town lock, stock, and wine barrel.

Today Los Olivos is pretty much all tasting rooms and restaurants up and down its crisscrossing main streets. The town has over thirty tasting rooms and counting. The filling station is now divided into a general store and, yes, a tasting room.

Across the main street from Fess Parker's Wine Country Inn sits the Los Olivos Wine Merchant & Café. The building once contained two entirely different stores, the Almond Vineyard Deli and an antique store, but when Sam Marmorstein bought the property in 1995, he took out the center wall and opened the place to both a fine dining restaurant and a wine store.

A wall of wine on the far side away from the restaurant, where bottles for sale lay in racks, attracted the attention of Jane Stewart and Alexander. They chose this as the location for the pivotal dining scene with all four actors. Renting the place for three days, the filmmakers rearranged the establishment so dining tables were moved out of the restaurant and set up directly in front of the wine wall.

The scene involving the four main actors is deceptive: on film, it's a simple dinner. But it's also a long scene that Alexander and Phedon need to keep visually interesting. What the writer-director wanted here was to show gracefully and effectively the progression of an evening, how the four characters get to know one another better, Miles—having just learned of his ex's remarriage—becoming inebriated, and how much this alarms Jack and impacts his possibility of having a successful, in his mind, evening.

The screenplay, for once, dictated that much of this would be improvised among the actors. Alexander knew that along with Rolfe Kent's music and Kevin Tent's editing, he would have to pull all these various fragments together in postproduction.

Alexander also, for once, relented on his no-makeup order to Virginia. His actress told him she needed mascara for this one scene.

Why? he wanted to know.

Because if a gal goes on a date, she's got to do a little something, she explained to her male director. And that something will be different from her other scenes. "There's got to be a little touch, maybe a necklace, something. He agreed with that," she recollects.

Cast and crew filmed the sequence over two nights with a room full of Alexander's "background actors," resulting in, as Kevin remarks, "a ton of footage." Phedon says he did "a lot of tight, floating shots," unlike any in his director's previous films. "We very much let the movie tell us how to shoot it."

The intent was for the scene to build momentum as the characters get caught up in the romance of the evening and Miles gets steadily drunker. Insert shots were made of wine and food arriving, Stephanie and Jack getting friendlier by the minute, Miles pontificating on wine, Jack reaching over the table to keep Miles from refilling his wineglass—a sort of "slow down, big guy."

"Paul is a fucking great actor," says Sandra. "In that scene, Miles is carrying all that emotional weight. Three characters are having a lovely time, eating and exploring, and the key character, Miles, is going through an emotional turmoil and Jack is keeping his eye on it. I remember telling Alexander, 'You gotta understand how lucky you are to have Paul Giamatti.' Sitting across from him at the table I could tell he is not only a beautifully instinctive actor, he has great technique."

If Paul himself has only vague memories of how he deployed his technique in that scene, there's good reason.

"Most of the time we drank fake wine, which was horrible. But the wine I was drinking—I don't remember what but Alexander was very specific about it—had to be real because of the color and consistency and legs it had. We couldn't fake it. So, I had to drink about two and a half bottles of the stuff. I got hammered shooting that scene. I was drunk."

The sequence ends when the women return from the restroom and Stephanie suggests everyone go to her place for more wine, cheese, and who knows what?

"The scene ends then," says Alexander. "Thomas Haden Church did about ten different improvs for how the scene could end—all very funny. He was extremely good that night—on point the whole time."

Thomas recalls, "We were close to wrapping for the night. Everyone started drinking. Alexander climbed up a ladder behind Paul and Virginia and wanted me to do whatever came into my head that moment after I ask for the check. We did many takes and because of all the extras in the restaurant I was trying to make everybody laugh. Why I meowed I don't have any idea. But every screening it gets a huge laugh. It's so ridiculous."

Good fortune smiled on all whose wine got mentioned or shown in *Sideways*. One example is the Fiddlehead Sauvignon Blanc from winemaker Kathy Joseph mentioned at the top of this dinner sequence. In his location scouting Alexander had visited Fiddlehead in the Lompoc Ghetto, as locals called the wineries that took over an industrial complex, and invited Kathy to a "movie night" at his rented house with pizza and wine.

She brought a wine case that included her Hunnysuckle Sauvignon Blanc, a limited bottling made in the Bordeaux style with oak aging and a creamy texture. In other words, this was a white wine made like a red wine. Alexander loved it so much that he told Kathy he was going to write the wine into the movie.

Fiddlehead Collars landed on the *Sideways* tour schedules. People as far away as Japan sought her out and sales zoomed.

Which is why Sam considers what happened to him during filming in his café a "calamity."

Michael, being a smart producer, tried to bargain down the price for closing his restaurant.

"You're more expensive than The Hitching Post," he complained to the restaurateur.

"Well, we do both lunch and dinner, so we're losing more," Sam replied.

"What about if we feature your wine?" asked Michael. "Can you take some money off?"

Like Frank, Sam made his own wine, called Bernat.

Sam ruefully explains, "At that time, I didn't think the movie was going to go anywhere, so I said, 'No, that's okay. I'll take the money.' That was a stupid decision. It was a mistake."

***

A line in Rex's novel during his dinner scene suggests a hallucinatory sequence. It comes when Miles realizes he is getting mightily drunk, and any moment he might "slip through a crack in the floorboards and find myself rowing across the River Styx with my demon entourage."

So Alexander and Jim have the screenplay shift from the café to the underworld—DARK AND TIMELESS in the scene heading—where Miles boards an open boat on an underground river with ghoulish human cargo and the hooded boatman Charon (played by casting director John Jackson) wielding his long staff. The movie was meant to cut between these two scenes, metaphysically so very far apart, as Miles walks to and from the pay phone.

The scene never made it into the movie.

"It looked good on paper," Alexander shrugs. "It didn't feel right."

His editor concurs: "Too many elements were going on where it was scripted that Paul's character gets drunk and calls his ex-wife. [The River Styx sequence] got in the way of the emotion of Paul's character collapsing, getting drunk and more desperate. When you threw that in there it was like—'What?'"

***

Another night shoot was the sequence in Stephanie's cottage. While Jack and Stephanie are mostly off-screen making the kind of noises that cause Maya to speculate that their friends are hitting it off, Miles and Maya get into a wine conversation rich in subtextual meaning. Each actor has a short soliloquy about wine that could be an actor's dream. Here the film pulls the viewer into the inner lives of its two potential lovers, who find intimacy painfully difficult to achieve as their romantic pasts haunt them.

A few weeks earlier during preproduction, Paul and Virginia had rehearsed the scene in Alexander's rented house. When they finished, Alexander congratulated them on the beauty of the scene, then gave his actress specific instructions.

"Now, Virginia," he told her, "I want you to put down your script and I don't want you to look at it until we're filming the scene."

She started to object, but Alexander interrupted her: "You already know it. So don't look at it."

By insisting she not look at the scene again before filming it, Alexander might have handed Virginia the Oscar nomination. Before shooting that pivotal scene Virginia went back to her house in Ventura for a few weeks while other scenes not involving her were shot. The screenplay, as Alexander directed, sat unopened.

"So, I'm looking at the script and it's looking at me and I tell myself, 'Okay, you know it.'"

Returning to the location, she did crack open the script to learn other scenes but remained determined to not look at *that* scene.

"It was sitting on a coffee table in my hotel room and I felt it was mocking me," she says. "I was so tempted. I wanted to open it and just *read* it. I was alone. It took a lot of discipline not to. Somehow Alexander knew that I would overwork it, that I'd think about it too much. He was absolutely right.

"When we got to the scene, the words were just there. That kind of writing doesn't take a lot of time to memorize. When writing is not good *that's* the bitch."

The shooting went well into the morning, but not because of multiple takes. "You don't need a whole lot of takes with those actors," says Alexander.

"I vividly remember the shooting of Virginia's wine speech," says the producer. "We shot all night at that location and the owner didn't want us all night. I had to run out to the edge of the property, where I was trying to negotiate with him to leave us alone and let us shoot. When I walked back in total darkness to that screened-in porch, I was listening to her speech on my headset at three in the morning and totally forgetting it was a movie—like I was eavesdropping on the most romantic, unexpected pairing of two human beings in the world. That was the high point of the shoot."

"Alexander creates this atmosphere where you feel like you're hanging out and he happens to have a camera on you," muses Paul. "We're hanging out and oh, by the way, we're making a movie. I remember feeling so relaxed and lovely and sweet [in the porch scene]. It didn't feel like a big deal. Literally, it was like she and I were having a good time."

***

Thomas was hesitant about the love scenes between Jack and Stephanie, since the woman he was boisterously romancing—"I went *deep* last night"—was his director's wife. He discovered his director held no such qualms about another man wooing his wife on camera.

"Alexander will push you and Sandra had a go-for-broke demeanor about it," says Thomas. "There is a naughtiness to Sandra."

For the scene in which Miles opens the door to their shared motel room and stumbles upon Jack and Stephanie in flagrante delicto, the action wasn't to Alexander's wishes.

"At one point Alexander said, 'You're making it too romantic. I want it to be more visceral.' So, we did more takes. 'I want it to be more raw,' he insisted. So, we did."

In another sequence, Jack and Stephanie slip away from a tedious wine lecture in a winery and find themselves in a barrel room for heavy-duty smooching. The crew was shooting up on a staircase where the actors were kissing, and Alexander was urging both actors to go for it.

"Let's do it," said Sandra.

"I was the hesitant one since this was the director's wife," Thomas says.

"The love scenes were just hilarious for me," remarks Sandra. "The funkier the better because that was like the characters."

The number-one question Sandra gets asked about her role in the movie is about the helmet-beating scene. This occurs after Stephanie learns that Jack, after all his intense wooing and promises made over several days, is getting hitched that weekend.

"She beat the shit out of me," says Thomas. "I turned down and away, which was off camera, but she hit my back and shoulder pretty good."

Both actors ended up aching from that beating.

"I was really sore the next day. I did not warm up well enough for that one," Sandra laughs.

"For me, the scene was like the beauty of comedy. Stephanie is crying through the whole scene. She is so hurt and furious. She is really emotional through the whole thing. I really feel for her."

The famous spit bucket sequence was filmed at the Fess Parker Winery. Because Miles will hugely deprecate the wine being sampled—"Tastes like the back of a fucking LA school bus"—Alexander renamed the location "Frass Canyon Winery." This is an inside joke that only an entomologist could love, as "frass" means insect poop.

Paul had to get Miles's wine bath right in a minimum of takes, though. "We did it in three takes because we only had three of those polo shirts," he muses.

The only accident came when a naked M. C. Gainey chased Paul toward his Saab after Paul had retrieved Thomas's wallet. As Gainey hit the car's driver's side, the glass broke.

"It was very scary," says Alexander. "It was on the first take that it happened. His gut hit the window at the exact moment when Paul closed the door and it shattered. We had to stop for two hours to replace the window."

***

Phedon would go on to shoot Alexander's next three features. Yet in the immediate aftermath of principal photography, he had mixed feelings about *Sideways*.

"I enjoyed the simplicity: a car pulls up in front of a house instead of a big camera move or dolly move," says Phedon. "Your pan just comes to a stop. You see the car and house—that's perfect. It tells the story. I'm a director myself now and I always think, what would Alexander do? I don't need to overproduce this. A shot can be simple and efficient."

And yet. . . .

"Being younger and a bit more ambitious and wanting to do too much, initially I was struggling with his *banalité*," he admits. "Ironically, I just rewatched *Sideways*. I had not seen it in ten years. Watching it reaffirmed for me it was perfectly shot—just the right amount. But back then, I don't know, I thought it looked a bit flat, not stylized enough with the lighting. My photography was not strong enough.

"I was at a screening of *Sideways* with Vilmos Zsigmond [DP on *Close Encounters of the Third Kind* and *The Deer Hunter*] sitting next to me. I said, 'It's too bright.' He said, 'No, it's a perfectly shot movie.' I thought he was just being nice. Twenty years later, I kinda agree."

Pinot Noir from Sonoma County, California

The vast Sonoma Valley wine region is home to nineteen AVAs, with sixty-plus grape varieties grown in the county. The key Pinot Noir AVAs are Sonoma Coast and Russian River Valley.

Sonoma Coast AVA

Highly coveted Pinot Noirs are produced in the extreme vineyards of this windy AVA, where some vineyards perch on ridgetops high above the Pacific. Within this large area lie the dramatic West Sonoma Coast, a fifty-one-mile sliver running along the Pacific, and Fort Ross-Seaview, with its rugged terrain jutting above the fog line.

Our suggested Pinots of Sonoma Coast:

Copain Wines Sealift and Cote Bannie
Flowers Vineyards Sea View Ridge
Fort Ross Vineyards Stagecoach Road
Hartford Family Winery Far Coast
Hirsch Vineyards San Andreas Fault and Estate
Kosta Browne Gaps Crown Vineyard (Petaluma Gap AVA)
Littorai Wines The Haven Vineyard
Marchelle van der Kamp (Sonoma Mountain AVA)
Naidu Estate
Nicholson Ranch Cactus Hill
Peay Vineyards Walala Vineyards
Vincent Christopher Walala Vineyards
Wayfarer Fort Ross Seaview Wayfarer Vineyards

Pinot Noir from Sonoma Coast, Sonoma County, California © ZW IMAGES

Pinot Noir from Russian River Valley, Sonoma County, California © ZW IMAGES

Russian River Valley AVA

One of the largest wine regions in California, Russian River Valley sprawls over 150 square miles, with over 1,400 acres under vine and home to ninety-three wineries. The AVA takes its name from its 110-mile-long river.

Nurtured by seasonal fog and cool marine air from the Pacific, the region's grapes enjoy an extended hang time.

Our suggested Pinots of Russian River Valley:

Charles Krug Dr. Maurice Galante Vineyard
DeLoach La Bienvenue and Estate
Eleven Eleven Bacigalupi Vineyard
En Route Les Pommiers
Gary Farrell Vineyards & Winery Hallberg Vineyard Dijon
    Clone Russian River Valley
Hartford Family Winery Jennifer's Vineyard
Kistler
La Crema Saralee's Vineyard
Lynmar Estate Bliss Block and Block 10
Merriam Dianna's Vineyard and Three Sons
Merry Edwards Winery Olivet Lane
Migration Running Creek Vineyard
Siduri Dutton Ranch

13

A Dew Drop at the End of a Leaf

Kevin Tent is another of Alexander's E Street Band to have graduated from the Roger Corman School of Low Budget Exploitation Filmmaking. Since graduation, he has edited every single feature for Alexander. He interviewed with Alexander for *Citizen Ruth*, showed a reel of his previous work, and won the job.

The two discovered that whatever minor disagreements they might have in the cutting room, they are always in sync over performances, meaning which take is the actor's most convincing.

For *Sideways* the editing took about nine months. The screenplay weighed in at 140 pages, which Alexander declares his longest to date.

"I knew I didn't want a 140-minute movie," he says. He also knew a screenplay so heavily dependent on dialogue gave him a falsely inflated page count. "We spent the most time on the dinner scene and split-screen sequences. Those took weeks to edit."

Once the River Styx scene came out of the dinner sequence, however, the path forward became much clearer.

"[Its removal] helped make the dinner scene better and superfluid with dissolves, and music, and montage," says Kevin. "The original editing had Miles go to the bathroom door and wait. One of our big breakthroughs in cutting that sequence was when we started to flash forward to him dialing the phone and then him getting up and walking drunk. The sound effect of the phone ringing faintly long before he started out—we stole that idea from *Once Upon a Time in America* [Sergio Leone's 1984 gangster film], where a phone is ringing a really long time and you couldn't know what the sound was doing there."

While editing *Citizen Ruth*, the two talked about favorite films and editorial tricks. One was the original *The Thomas Crown Affair* (1968), where director Norman Jewison and editor Hal Ashby did split screens to show simultaneous actions, such as the Boston bank robbery that opens the film.

For *Election*, they managed a tiny moment of split screens. Then while *Sideways* was in production, Kevin was getting dailies of second-unit shots of an ostrich staring at the camera and farm workers picking grapes. He called the director on the set at lunchtime to ask where this footage was meant to go in the movie.

"Oh, I thought maybe we'd do a split-screen sequence," said Alexander, reviving their long quest to significantly tackle split screens.

"It worked out great," says Kevin. "The split screens jump the story and help save a lot of time. The shot of an ostrich sets up Jack's line about running naked through an ostrich farm. Split screens are hard to do—but fun."

The split screens also give *Sideways* lively visuals without the frequent cutting so many ADHD-afflicted films do today.

"We make a concerted effort to try not to cut unless we really have to," says Kevin. "We try to stay in a shot longer."

The rough cut came in at about two hours and fifty minutes. Then the pair had another go at it. They also, as they always do, screened the hell out of the movie for friends and family.

"We screen constantly," Alexander says. "Movies for me [are] a long time in prep and a long time in post. Shooting, the most expensive part, should go like clockwork."

One drawback to this lengthy editing process was that he grew to doubt the wisdom of the dueling wine speeches, a sequence most of the movie's admirers would see as the heart and soul of the film. When a director, especially a writer-director, hears the same lines over and over while editing, he can make the mistake of cutting them out because they sound dull by this point.

"He thought it was too sentimental," says Rex. "'We've got to cut that speech,' he kept saying. The sound mixer was saying, 'Are you kidding me? It's very moving.' 'No, sentimentality dates movies!' Alexander said."

"I thought it was too talky," Alexander admits. But he did show each new cut to filmmakers he respected.

"I remember Mike Nichols telling me 'I was enjoying the film but when they sat together on the porch and started talking about wine, that's when I really sank into the movie,'" Alexander recalls. After Nichols voiced his strong approval Alexander realized that the speech "plants the flag in a deeper emotion. It lends the film some very enjoyable gravitas."

The speech stayed in.

***

Alexander Payne on set with Thomas Haden Church and Paul Giamatti PHOTO BY EVAN ENDICOTT

Alexander Payne rehearsing a scene with Thomas Haden Church and Paul Giamatti. A crew member passing by PHOTO BY EVAN ENDICOTT

Alexander Payne and Paul Giamatti
PHOTO BY EVAN ENDICOTT

Cast and crew wrap party at the Windmill Inn later renamed Sideways Inn PHOTO BY RACHEL FLEISCHER

The musical influence on *Sideways* comes from *Big Deal on Madonna Street* (1958), Mario Monicelli's farce about the attempted burglary of a pawn shop by inept thieves. Alexander loves Piero Umiliani's jazz score for the movie. Again, he was striving to combine that era's Italian comedy with the sensibility of a seventies American movie.

So the musical vibe sought was from the mono past. What Rolfe Kent settled on to capture period sound was vintage equipment. The English-born composer's working relationship with Alexander goes back to those short erotic films done by Playboy Video Enterprises and Propaganda Films.

"We have a friendship, but it doesn't necessarily get any easier," the composer says. "We're always pushing to see what else we can discover. *Sideways* was particularly difficult for me because it was jazz and I didn't have a handle on that because so much of composing is on computers now, which does not lend itself to jazz. Great musicians are where jazz comes from."

Rolfe's engineer Greg Townley studied past recording techniques, from three-mic mono to all sorts of ways Motown and vintage blues were recorded.

"One thing I learned is when you want the sound of a John Coltrane saxophone, it's not a pristine digital recording you want, you want a vintage rhythm microphone—you want it saturated, overblown," he explains. "So even though the movie was recorded digitally, it was recorded through 1960s equipment, which were Bakelite controls and microphones, all vintage."

Normally, in a music scoring session for a feature, the conductor and musicians must fit the music with exactness to the already cut film. But when it came to *Sideways*, because it was jazz musicians getting into the groove, Alexander didn't worry so much about how the music would fit the picture.

"He said, 'Let's make it as great as you can as music, and then we'll figure out how it fits together.'"

Film composers are usually asked to do a synthetic "mock-up"—a demo—so everyone feels safe going into the actual recording session. The problem with jazz is that jazz doesn't mock up in that way.

So the scoring for this movie was highly unusual in that Rolfe had three recording sessions spread out over time so he could revise the music and course correct when necessary.

"By the time of the final session the LA musicians we had were absolutely the best musicians," he notes. "They absolutely nailed what we wanted and added things we didn't expect."

And yet for the crucial Los Olivos dinner sequence, "the bulk of it is simply me recording in my studio and humming and playing lots of instruments. That's almost a dream sequence as Miles gets drunker and drunker, which Greg enhanced no end by his mixing. There were some wind instruments in there, but a lot of humming."

The Miles-and-Maya love theme Alexander establishes is a mid-1960s piece, "Symbiosis," by the influential American jazz musician and composer Bill Evans. Alexander and music editor Richard Ford establish it during the porch wine discussion; reprise it when the two go to her apartment and become intimate, then again when they visit the farmers' market the following day and at the very end, when Maya leaves a message on Miles's answering machine.

"Rolfe wanted to match it so the movie's score would be all his, and he came up with a lovely piece that could not match the emotions, I thought, of the Bill Evans piece," says Alexander.

Says Rolfe with a laugh, "*My* Miles-and-Maya love theme is on the soundtrack album if you want to hear it."

***

One delicate bit of ADR—automatic dialogue replacement, or looping—was needed in postproduction, and that was Maya's voice message on Miles's answering machine at the movie's ultimately hopeful, quiet, and possibly romantic ending.

"We tried the ADR phone call a few times," recalls Virginia, "but I thought this needs to be on a phone. I asked Alexander, 'Is there a way for me to call it in?' Alexander said, 'That's a good idea.' So he set me up in a different sound booth with a phone where I could just be alone and read it there and he recorded the sound. It worked!"

This ending is much remarked on by people who want the filmmakers to explain what happened following Miles's knock on her door.

"I believe the ending of the film belongs to you, the audience," says Virginia. "Do they begin their lives together, or maybe she wasn't there? Most people I talk to think it's a romantic ending."

Certainly, costume designer Wendy Chuck put an optimistic orange shirt on Paul Giamatti, the first time in the movie Miles wears a brightly colored shirt instead of his usual shades of blue or maroon. Orange, Virginia points out, is the color of courage.

"In my opinion, they sit down, have a cup of tea, and begin their lives together," says Virginia. "Paul would say, 'Naw, these people are drinking too much. They'll never work.' Thomas would say, 'My character was found shot to death in an alley in Bakersfield.'"

Paul, when interviewed, takes a more nuanced approach: "That guy really needed to get his shit together." He pauses. "Maybe she would help him get his shit together."

Sandra imagines her character "is still living up there and helps manage a tchotchke shop."

We'll leave the final word for Alexander.

"People ask me, 'What happened when she opens the door?' I say she didn't open the door. A family of Vietnamese immigrants moved in there and have no idea where she's gone.

"It doesn't matter. The story is him. You want the end of some movies to have a dew drop at the end of a leaf. Is it going to fall? Maybe it just starts to fall—and you cut."

Pinot Noir from Monterey County

Framed by the Gabilan and Santa Lucia Mountain ranges, Monterey County's weather pattern varies dramatically from the cooler northern part to the warmer southern end. Among the county's ten sub-AVAs, Monterey is the largest, with forty thousand acres under vine. While all the AVAs reflect their distinctive characteristics, the Santa Lucia Highlands (SLH) is the jewel in the crown and a prime Pinot and Chardonnay zone.

Santa Lucia Highlands

The unique windswept environment, settled by Swiss-Italian farmers in the previous century, allows the SLH fruit a longer hang time on the vines, thus deepening the flavor in the grapes. Some fifty vineyards perched atop terraces of the namesake mountain ridge covering twenty-two thousand acres are tucked around mesas scaling up to 2,300 feet in elevation. Wineries are few, lined along the River Road Trail, and many are not open to the public.

The SLH fruit has an iconic status, as West Coast winemakers crave access to fruit from such pedigreed vineyards as Pisoni, Rosella's, Sierra Mar, Hahn Hook, Tondré, McIntyre, and Boekenoogen.

Our suggested Pinots from SLH AVA:

Clarice Rosella's Vineyard and Garys' Vineyard
Lucienne Doctor's Vineyard
Morgan Double L Ranch
Pisoni
Roar Rosella's Vineyard
Talbott Fidelity and Sleepy Hollow Vineyard
Tondré Wines
Tudor Wines
Wrath McIntyre Vineyard

Our suggested Pinots from other Monterey AVAs:

Albatross Ridge Vivienne's Cuvée and Estate Reserve (Monterey County)
Chalone Vineyard Estate and Reserve (Chalone AVA)
J. Lohr Fog's Reach (Arroyo Seco AVA)

San Benito County, California

Next door to Monterey this county is where the late Josh Jensen established Calera, which some call "America's DRC." It's one of the country's great wines.

Calera Wine Company Jensen Vineyard (Mt. Harlan AVA)

Pinot Noir from Monterey County, California © ZW IMAGES

14

We've Got Gold

The first public screening of *Sideways* happened in Santa Barbara's Arlington Theater. The Spanish Colonial–style cinema, which opened on State Street in 1931, gives audiences the impression they're sitting outside in the plaza of a colonial town. The walls feature houses, staircases, and balconies, not painted but built out from the walls. It's one of Alexander's favorite movie theaters.

Guests included many of the wine region's vintners, restaurateurs, and others who helped make the movie possible. The reaction was strong, especially from those who understood what this could mean for their region and businesses.

"We were shell-shocked," says Frank. "I don't think I've ever seen a movie that promoted a business so organically and so well and made it almost a character in the movie."

The movie also changed his mind about that yellow Hitching Post sign, which had been sitting for years in front of the restaurant on Highway 246.

"We always hated the color and wanted to change it. When we walked out of the theater, I said, 'That sign needs to be on the front page of our website.'"

Fox Searchlight executives felt they had a possible art-house hit that might expand into mainstream cinemas if it got award nominations. Alexander thought he had made "a nice, slight little story." Paul was more pessimistic: "Who the hell is going to care about a movie about wine nerds?"

Alexander and Steve Gilula went about securing a festival/platform release. This meant that *Sideways* would play the fall film festivals before its October launch in four theaters, with the hope that reviews and word of mouth would support its gradual expansion into more cities over the coming weeks.

The marketing team worked long and hard to come up with an image that fit an unusual midlife-crisis comedy set in wine tasting rooms, vineyards, wineries, and restaurants. The final image—a cartoony sketch of two puzzled men trapped inside a wine bottle lying on its side against a green backdrop—flummoxed studio executives.

"Certainly, to Fox senior management that was not considered a commercial image," says Steve. "It was very artistic and in the tradition of European posters. Alexander liked it. It didn't fit with the new trends, but it was very distinctive and fit the film."

Steve, who had cofounded the Landmark Theater chain with a single theater, the Nuart, in West LA in 1974, helped build Landmark into the first successful national art circuit. For nearly twenty-five years he traveled the country, meeting exhibitors and visiting cinemas in what most of the movie industry considered flyover country.

"I learned that in the latter part of the twentieth century, local newspapers and film critics had tremendous influence and power. Cities like Denver, San Diego, and Minneapolis were incredibly vital, dynamic theater markets but overlooked by distributors in LA and New York."

Before Steve joined Fox Searchlight, the typical release pattern for specialty or art films would be to open in New York and LA, see what the reviews were like, and then decide whether to spend any more money to go out to the rest of the country.

"That's not very constructive," says Steve. "I said if we have a good movie and it doesn't do great in New York or Los Angeles, there are theaters around the country that will still play the movie. Set the movie, send one-sheets, send trailers, and play the movie.

"On *Sideways* we had the whole country set before it ever opened. We gave our films a profile as opposed to relying on a *New York Times* review."

***

Initially, the festival selection committees were lukewarm on the film, which was a surprise. It did not premiere in Telluride, the first fall North American festival, nor in any of the European festivals. Fox Searchlight did bag a world premiere at the Toronto International Film Festival and then an American premiere in the New York Film Festival's prestigious closing night slot. So expectations were modest, because the first litmus test for a fall release is whether or not festivals are fighting over a film. That didn't happen.

Sideways premiered on September 13, 2004, in Toronto's Elgin Theatre, a cavernous 2,100-seat theater, home to vaudeville acts and movies since 1913. The film had a morning press screening a day earlier at the Varsity multiplex in Yorkville. So ecstatic reviews were popping up online even as festivalgoers were lining up along Yonge Street.

The filmmakers and cast came out on the old stage before the screening and got sustained applause. Alexander introduced the cast, including his wife, the "queen of the Toronto Film Festival," and soon the house lights darkened.

When the lights came up two hours and seven minutes later, the audience jumped to its feet, turned toward the seated filmmakers, and roared their approval.

"You can never anticipate that moment. You can never repeat that moment," says the producer. "This was the first time for most of us that that ever had happened."

The reviews floored everyone—the filmmakers, Fox Searchlight, and hungry exhibitors.

Roger Ebert thought the movie "adds up to the best human comedy of the year—comedy, because it is funny, and human, because it is surprisingly moving."

The *Village Voice*'s J. Hoberman opined, "Jack and Miles are male archetypes, as well as the two most fully realized comic creations in recent American movies." He also prophesized that "the movie should consign merlot to the bargain rack while, thanks to Miles's showstopping disquisition, sending pinot noir orders through the roof."

The New York Times's Manohla Dargis cited the scene where the actors ruminate over wine as a "small masterpiece, this exquisitely shaped scene shows just how far Mr. Payne has come as a director, especially of actors. It took courage to cast Mr. Giamatti in the central role, not because he isn't up to the challenge, but because he's neither pretty nor a star, two no-no's in the contemporary film industry."

"Those were some of the best reviews Fox Searchlight ever had," says Steve.

From the initial four theaters *Sideways* opened in on October 22, 2004, the film, after Oscar nominations were announced, eventually played in close to 1,800 North American theaters, an unlikely if not implausible number for this kind of movie.

The film soon entered "awards season," which happens each fall and carries on through the Academy Awards ceremony, which in this instance took place on February 27, 2005, at the Kodak Theatre in Hollywood. Major critics groups such as the Los Angeles Film Critics Association (a sweep across the boards), the New York Film Critics Circle, the London Critics Circle Film Awards, the Chicago Film Critics Association (another sweep), and the National Society of Film Critics, along with critic groups all around North America and several international groups, bestowed nominations if not awards on the film.

The LA Film Critics Association's banquet, which took place at the InterContinental Los Angeles Century City Hotel, looked like a Santa Barbara Vintners Association's Harvest Festival. The Hitching Post and quite a few other members hosted tables where their wines were poured at a predinner reception. During the dinner, Frank snuck over to Alexander's table and gave him a signed magnum of Hitching Post Highliner Pinot Noir.

"I realized at a certain point we're going to get a lot of awards attention because the movie is about a writer and movie critics are first and foremost writers," says Michael. "They are seeing themselves and seeing their own struggles, ups and downs, and we're going to kill in the awards race.

"I don't mean that cynically, but writers were the best possible audience for a movie about a thwarted writer and the struggles of writing. It was a huge blessing we had all these advocates and champions who see themselves in Miles. The casting choice helped, that it was Paul Giamatti instead of George Clooney. They wouldn't have seen themselves in Clooney as easily."

The Hollywood guilds, such as the Writers Guild of America (Best Adapted Screenplay), the Screen Actors Guild (Outstanding Performance by a Cast), and the Casting Society of America (Best Feature Film Casting) offered their kudos. The film also received nominations from other guilds, including the Directors Guild of America.

The Hollywood Foreign Press's Golden Globes went to *Sideways* for Best Screenplay and Best Motion Picture–Musical or Comedy. And let's not forget the Film

Cast Halloween party, Virginia Madsen, Paul Giamatti, and Sandra Oh   PHOTO BY EVAN ENDICOTT

Independent Spirit Awards, where the film swept every nominated category—best feature, director, screenplay, male lead, supporting female, and supporting male.

According to IMDb, *Sideways* earned 122 awards around the world.

***

These awards seasons, covered if not overcovered by the mainstream press along with the Hollywood trades, bloggers, and television entertainment programs, are as exciting and glamorous as they are debilitating and unnerving for actors and filmmakers.

As the season grinds on, cocktail parties and appearances for Q&As after screenings turn into seemingly endless award shows on the weekends, necessitating black tie for the men and new and dear outfits each evening for the women.

"For the rest of my career I've been reconciled to the *value* of the awards season," says Steve. "For *Sideways* it meant tens of millions of dollars at the box office. It stayed in theaters and allowed people to see it."

But for actors and filmmakers, it was a double-edged sword. For Thomas, this was a level of acceptance he was unused to, and he relished the opportunity to meet major people.

"At that point, I'd been fifteen years in the industry, mostly TV," he points out. "I had few hit films but never come close to that level of acceptance by critics, audience, industry peers. I met people wanting to work with me—Clint Eastwood and David Mamet and Ed Zwick."

Yet the grind of putting on monkey suits for lavish Beverly Hilton dinners and making small talk with the same group of hopefuls night after night can wear one down.

"You're in this strange bubble where every night there's an awards event you have to attend and every weekend another black-tie gathering of people," recalls Michael. "It's a business. It is an important part of the sales process. Alexander said, 'I just want to think about making another movie.' It detaches you from what you love most. Every day in Santa Ynez was like the best day of my life. That was the highlight."

Of the awards season, Alexander admits, "It sounds awful to say 'burnt out,' but we did get a little burnt out."

The Academy Award nominations themselves delivered a shocking blow to the *Sideways* team and Fox Searchlight. After being nominated and usually winning for Best Actor in nearly every competition throughout the season, Paul was not even nominated.

"Paul not getting nominated was the most heartbreaking thing," says Peter Rice. "By the time of the nominations, Paul had won almost everything. Here was a perfect gem of a movie and he's central figure. It felt so unfair. It was certainly a hugely important movie in the development of Searchlight."

While Michael feels that in 2005 Oscar voters were not yet focused on the world of specialty or indie films, he also notes that "Paul was an unconventional leading man."

The industry wisdom about the Academy's shocking snub of Paul is that Clint Eastwood's late entry in the Oscar derby with *Million Dollar Baby*, as both its director and male star, edged Paul out of the fifth slot for the acting nomination.

"We were the front-runner in the Academy race until the Clint Eastwood movie knocked the legs out from under us," says Steve. "Warner Bros. did a December drop and Clint Eastwood is Clint Eastwood."

"On the list of bad things happening in my life, that doesn't even make the list," says Paul. "How can I be disappointed in something I never expected to happen? It was wonderful and crazy that I was even in that movie at all."

He also had another reason to put the Oscar snub into perspective. His mother died right before the film's release. So, at best, the awards season was only "fun-ish" to him. The cocktail receptions, Q&As, and monkey suits were "only a distraction from having to deal with that. The whole thing to me was like, 'Sure, whatever.'"

In the end, all *Sideways* earned on Oscar night was a win, after one previous nomination in that category, for Alexander and Jim for Best Adapted Screenplay. This was what Academy voters often did in those days with oddball (in their minds) but fabulously successful films such as *Sideways* and *Pulp Fiction*.

***

"Walking into parties like *Vanity Fair*, you're expected to hold the Oscar all night," says an amused Alexander. "As you're walking down the red carpets, you hear people on walkie-talkies saying, 'We've got gold.'"

Alexander has long understood the true value of gold.

"When Jim and I got nominated for *Election*, my dad called to congratulate me. 'Now don't let this go to your head,' he said. 'No, but I want it to go to other people's heads,' I told him. You can get more stuff done."

There is also a value in *not winning*, too, as Virginia sees it.

"Things would have been different if I had won," she states. "There would have been an expectation for me to repeat myself. There was no way of repeating that. Then people would say, 'Yeah, I guess that was an anomaly.' But I got to keep doing what I do without the pressure of having won.

"I felt the night of the Oscars I represented all the actors who are treading water, working their way up, working their ass off for recognition, and not getting it. Going to the big dance, going all the way to Academy Awards—our film got to go."

The awards season, in the end, put *Sideways* and Fox Searchlight on the map.

"We were able to hold [in theaters] and then expanded," says Steve. "It's one of biggest specialty films of all time in the US—that's a big number. It defied logic."

The gross worldwide, again according to IMDb, was $109,706,931.

Sideways put Searchlight into the Best Picture game. It was the first of eighteen Best Picture nominations out of the 167 films Steve and Nancy would oversee during their tenure, which ended in April 2021 when they stepped down as cochairs after Disney bought out Rupert Murdoch's Fox film interests.

Peter Rice, along with others associated with the film that did wonders for The Hitching Post, received a magnum of Pinot with a label of the now-famous poster from the movie. "Frank Ostini said it was a 'bottomless bottle of wine.' Any time I've drunk it, I am to send it back and they'll refill it."

Streaming has turned the independent film business—and many indie filmmakers think the term is now finally obsolete—and the awards business on their heads, because the streamer can pay outrageous amounts that make no financial sense for an indie distributor such as A24 or even Searchlight Pictures.

Which is not to say that *Sideways* couldn't get made today, but it would be that much harder. The filmmakers would probably have to go to Netflix.

"Peter said, 'I'll roll the dice and believe in that filmmaker and that script,'" asserts Michael. "He had the power to do that because it was a time when heads of divisions like Searchlight didn't have their decisions micromanaged from above. He was able to just say yes. Probably that glass of Pinot helped push him over the edge."

15

The Oregon Trail

Oregon is Pinot country. Has been right from the beginning. In the story of the wine renaissance in post-Prohibition America, recovery began as early as the 1930s in Napa Valley. But as far as our book's protagonist is concerned, the Pinot Noir grape struggled in Napa's warm climate, only finding suitable weather in Los Carneros, down at Napa's southern tip.

Essentially the Pinot Noir success story in America begins in Oregon, not Santa Barbara. While the wine business regenerated more or less simultaneously in both areas—two UC Davis grads planted the first wine grapes in Santa Barbara County in 1964, while David Lett planted his grapes near Dundee, Oregon, in 1966—the key difference is that Lett planted Pinot Noir. Thus, wine historians dub him "Papa Pinot."

California growers planted all sorts of grapes, and it took more than a while to figure out that Burgundian varieties were the answer in cool climates, not warm. This bafflement happened throughout the state.

In Mendocino's Anderson Valley, nowadays lush with its mainstay Pinot Noir, Zac Robinson, owner of Husch Vineyards, recollects, "We knew other people planted Chardonnay and Gewürztraminer, but no one could figure out the red grape."

However, the prescient pioneer of Oregon Pinot, David Lett, knew exactly what grape was right for the Willamette Valley's soils and climate.

"We looked at *Sideways* kind of quizzically," muses his son Jason Lett, who took over Eyrie Vineyards in 2008 when his father passed on. "First of all, it's centered on California Pinot Noir, and California Pinot Noir, to us, is a newcomer to the scene. A lot of spots in California that are now so much about their Pinot Noir didn't get started until the late seventies and mid-eighties. They came at least ten years after us."

Why did the Oregonians, mostly young couples transplanted from elsewhere, so easily identify the state as Pinot country?

Well, the state can thank Lett and the now nearly forgotten Charles Coury, who arrived there in the 1960s. A trained meteorologist, Coury entered UC Davis in 1961 to earn a master's in viticulture. His master's thesis "Wine Grape Adaptation in the Napa Valley" suggested that cooler climates were better for growing certain grape varieties. He moved to Oregon in 1964 and began to explore the region. He looked for land, but he also visited the archives at Oregon State University to uncover the grapes and farming techniques utilized by pre-Prohibition farmers.

Beating him to the punch, however, was David Lett, who planted Pinot Noir (among many other varieties) in the red hills near Dundee. He too went to Davis, which had no winemaking program then, so like Coury he pursued a degree in viti-culture. The eight students in the graduating class of 1964 had formed a wine-tasting group, where everyone noticed a vast difference in tasting Burgundies and domestic Pinot Noir. Their professors explained why—climate. In other words, the same cul-prit at the crux of Coury's paper.

"Any fruit, whether an apple or grape, is going to achieve its maximum flavor where it can find the longest growing season," says David's son. "So if heat is pushing ripening to happen so quickly that you don't have a long growing season, then you're not going to capture the flavor. You must find those places where flavor expression can reach its peak before winter comes. Only three places my father identified out-side of Burgundy were going to work [for Pinot]: the South Island of New Zealand, which we now know would work; the north coast of Portugal near the Atlantic; and Willamette Valley."

David came to Oregon in 1965 with three thousand grape cuttings. He planted the vines in a nursery he established in Corvallis while searching for the perfect place. He and his wife Diana found it in an abandoned prune orchard and the fol-lowing year transplanted the cuttings to the Eyrie Vineyards. They discovered that a tree atop the vineyard contained a hawk's nest or an "eyrie."

"They felt a sense of kinship with that nest, so the name," says Jason. "My mother is an English major and knows a lot of obscure words no one can pronounce."

Willamette Valley, Oregon, is at nearly the same latitude as Burgundy, France, the motherland of Pinot. Willamette is located along the 45th parallel (which is more in line with Bordeaux), while Burgundy is along the 47th. As in Burgundy, Oregon winters are cool and wet and summers are—perhaps, as we shall see, *were* is a more appropriate verb—temperate with cool evenings. Yet the valley's soils are very acidic, versus the limestone and low-acidity soils in Burgundy. Most vineyards were planted on hillsides above two hundred feet, as the valley floor is too fertile. Wine grapes like to struggle in difficult soils.

Willamette Valley is today the heart of the state's wine industry, although good Pinot Noir is grown in AVAs such as Rogue, Umpqua, and Applegate valleys in the southern part of the state. Twenty to forty miles wide and 120 miles long, the Willamette Valley is a long, level, alluvial plain with scattered groups of low basalt hills. Oregon's largest AVA, from the Columbia River in the north to the city of Eugene in the south, was established in December 1983.

At that point, the Willamette was still an old farming community, with more hazelnut orchards and Christmas tree farms than vineyards. Everyone would say you could fit the entire Oregon wine industry in the back of Nick's, a McMinnville Italian restaurant where many winemakers hung out.

One pioneer, Susan Sokol Blosser, remembers even smaller confines for that industry in its earliest days: "There couldn't be more than five or six couples in the whole wine industry. We would meet in our living room. Charles Coury's paper was instrumental in our thinking."

David and Diana Lett, who had been joined in the quest for New World Pinot by Bill and Susan, Dick and Nancy Ponzi, and Dick and Kina Erath almost from the outset, got additional company when David and Ginny Adelsheim, Jim and Loie Maresh, and Joe and Pat Campbell came to Oregon to begin putting grapevines into the soils.

"We were underdogs," Susan explains. "We for the most part were liberal arts graduates. We knew how to work together; we knew how to do research. If we'd all been business students, we probably would have been much more competitive and not as collaborative."

That continues today. At their Chardonnay technical tastings, which began in 2015, over 140 winemakers and vineyard managers as well as vineyard and production staff gather in a room for a day to taste through eighty to one hundred barrel samples and lay out the whole process and data behind each wine. In 2023, the group launched the same format with Pinot Noir.

"You will not see that in a lot of wine regions," notes Alexana Winery's winemaker Tres Burns.

Another key point about Oregon is its French connection. David Adelsheim has been going to Burgundy for the past fifty years.

"He's sort of an ambassador for our region," says Gina Hennen, the winemaker for Adelsheim. "There's been a very long-standing connection to Burgundy with this area over the decades—people going back and forth and interns through the vintages. We tend to look in that direction more so than our neighbors to the south."

"I go to Burgundy almost every year, and when I do, it's amazing the interest there in Oregon," says Doug Tunnell, co-owner and winemaker at Brick House, who left network news reporting when his contract expired at CBS in Paris in 1992.

Indeed, it was French investment in Willamette Valley that first encouraged Doug to return to his home state. By 1975 David Lett identified a block on his property producing something different and intriguing. His Eyrie Vineyards 1975 South Block Reserve Pinot Noir got submitted to a blind tasting at the *Gault-Millau* Wine Olympiad in Paris in 1979. It placed third.

One of Burgundy's leading vintners, Robert Drouhin, head of the Burgundy *négociant* Maison Joseph Drouhin, was intrigued. He restaged the blind tasting the following year, but this time replaced the Burgundy entries with wines from his house. Eyrie finished second by a single vote.

When Robert's daughter Véronique, having taken a master's in enology in 1986, wanted to do a harvest internship in the New World, she naturally thought about California. Her father told her she should go to Oregon. It was more interesting, he told her. She followed that advice, and her father came out to see her. He took a look around.

The very next year Maison Joseph Drouhin bought land to establish a winery in Willamette. The wine world was shocked: a Burgundian house bought land for a winery in America!

"Burgundians did not buy land outside of Burgundy—they didn't need to," notes David Millman, president and CEO of Domaine Drouhin Oregon. "You harvest at the same time of year. They also had nothing here, no equipment. Just dirt."

Doug's reaction says it all: "Living in France at the time, I thought if the French are buying land here something worthwhile is going on." He moved back home.

Wine writers started to visit. Robert M. Parker Jr., the most influential American wine critic, went so far as to establish a partnership with his brother-in-law Michael Etzel and a French-Canadian investor for an Oregon winery called Beaux Frères (French for brothers-in-law).

Late in 1985, an informal group of Oregon winemakers and restaurateurs envisioned a premiere event in the heart of their wine country celebrating their best-known grape. Each year since the first event held in 1987, the International Pinot Noir Celebration (IPNC) brings wine lovers, chefs, journalists, and foodies to McMinnville.

International, mind you, and not held in Burgundy or California or New Zealand. But in Willamette.

***

Pinot Noir from Willamette Valley, Oregon AUTHOR'S COLLECTION

Pinot Noir, Willamette Valley, Oregon AUTHOR'S COLLECTION

Vintner Michael Etzel at his Sequitur Wine estate in Newberg, Willamette Valley, Oregon AUTHOR'S COLLECTION

As the authors wound their way along the Oregon wine trail to talk with vintners, two statistics continually cropped up. Oregon makes only 1.5 percent of all wine produced in the United States. Yet it produces over 20 percent of all *Wine Spectator* wine scores over ninety points, which is that magazine's demarcation line for excellence. Wine buffs might scoff at scores from critics whose methodology is vague and arbitrary.

"*Wine Spectator* scores are not an arbiter of quality," admits Jessica Mozeico of Et Fille Wines, "but it can be an indicator that we're producing a much higher quality from a much smaller portion [of the domestic market]."

So there is an outsized ratio of quality to production. Oregon can boast of having among the best quality Pinot Noirs in America, but they often come at a high price point.

"Expensive Pinot Noir is what we make," admits Susan. "So, over the years, Pinot Noir has been a tough sell."

Things have changed, though. For one thing, vintners have figured out how to get a twenty-dollar Oregon Pinot into the marketplace. By using machine harvesting, tank fermentation, and micro-oxygenation with wood inserts rather than aging the wine in a barrel, winemakers can introduce a pretty good Oregon Pinot Noir to a new customer.

Climate change has also helped Oregon get to that benchmark. "With warmer weather and densities of plantings we can make great Pinot Noir at twenty dollars a bottle," insists Adelsheim's general manager, Rob Alstrin.

Oregon is still mostly about small producers making five thousand cases or less of wine. Now, however, large wineries such as King Estate, Four Graces, A to Z, and Acrobat can offer more modestly priced Pinots through sheer volume.

Another thing happened—finally. It took a while after the successful Drouhin experiment, but outside investment is rampant in Oregon, coming from France, Italy, and California, as Jackson Family Wines now owns such Willamette jewels as Penner-Ash, Gran Moraine, and WillaKenzie.

Oregon Pinots are getting on restaurant wine lists. When Jessica, who cofounded Et Fille with her late father in 2003, started, she had to explain why Willamette Valley Pinot belonged on a list.

"Restaurants would say, 'Oh, I've got a lot of Pinot on my list. I'm all set,' meaning 'I've got a lot of California Pinot,'" she remembers. "There are very different conversations now. Most higher-end restaurants know they need to have a Willamette Pinot Noir on the list."

***

When the authors toured Willamette a quarter century ago, the Pinot Noirs were very good but also very pale, forest-y, ethereal yet not fruit-forward. Vintners were then looking for ways to preserve color, because consumers often think Pinot Noir is a lower-quality grape because of its lighter color. On this visit, the Pinots were equally complex but darker and richer, more red cherry to black cherry to plum.

Sure, farming techniques change and vineyards grow older and better, but the main factor seems to be climate change and extreme weather events.

"Twenty-five years ago, we were struggling to get ripeness and enough sugar to make alcohol," says Doug. "Now, for the last ten years, it's the opposite. We have to reduce exposure to sun and the impact of ripening rather than enhance it."

Michael Etzel puts it this way: "We used to focus on sugar; now we focus on acid and pH."

The year 2011 was the last cool vintage of that decade. From then to 2015, things got warmer and warmer, with 2014 the first of the hot vintages. In 2020 forest fires filled the valley with smoke and nine days of apocalyptic red skies. The crop was mostly lost.

The year 2021 brought a heat dome in June that reached an unprecedented 113 degrees. "The vines had never experienced such heat," says Tres. "Fortunately, it was early enough in the year that we lost only a few vines."

The viticultural response to climate change has been to plant on northern slopes, which are now much warmer, and higher on the southern ones. "When I first started we planted the southern slope at less than six hundred feet," says Michael. "Now we plant [at] one thousand feet to get cooler climate conditions."

Other changes include leaving more fruit on the vine to slow down ripening and grafting California clones to better deal with the heat. Or simply picking earlier. At some point, though, if the climate changes significantly in the world's great wine regions, phenolic ripeness may not have happened by the time you want to pick grapes. "So that's when you start running into trouble," says a concerned David Adelsheim.

Because then you might have to change your plants. "If we're not known for Pinot Noir anymore," says David, "what then? That's a huge and very difficult story, but no different than Napa and Cabernet. Every place in the world is dealing with the same issue—[mitigating] the problem [of climate change] so they can continue to be known for the same wine in a style not too far off from what people expect."

"We're farming on the edge of success," says Melissa Burr, VP of winemaking at the Stoller Wine Group. "Because we have rain, we are always looking at risk and botrytis mold. You have to farm with the worst-case scenario in mind."

That's Oregon in a nutshell: Farming on the edge of success in the high-risk, low-yield farmlands of Willamette Valley.

Definitely Pinot country.

Pinot Noir from Willamette Valley, Oregon

Oregon's Willamette Valley is the heart of the state's wine industry. Pinot Noir is king, followed by Pinot Gris, Chardonnay, and Riesling.

From some fifty wineries when the AVA was established in 1983, the region has expanded to over seven hundred wineries spread throughout eleven nested AVAs. Chehalem Mountains, Dundee Hills, Eola-Amity Hills, McMinnville, Yamhill-Carlton, Ribbon Ridge, and Van Duzer Corridor are among the key appellations.

Our suggested Pinots of Willamette Valley:

Adelsheim Ribbon Springs and Breaking Ground (Chehalem Mountains District)
Alexana Single Clone Pommard (Dundee Hills District) and Fennwood Vineyard
    (Yamhill-Carlton District)
Beaux Frères (Ribbon Ridge District)
Brick House Evelyn's (Newberg District)
Chehalem (Chehalem Mountains District)
Domaine Drouhin Laurène and Rose Rock (Dundee Hills District)
Domaine Serene Fleur de Lis Vineyard (Dundee Hills District)

Pinot Noir from Willamette Valley, Oregon © ZW IMAGES

Et Fille Wines Fairsing Vineyard (Yamhill-Carlton District)

Evening Land La Source and Summum (Eola-Amity Hills District)

Eyrie Vineyards The Eyrie (Dundee Hills District)

Fiddlehead Cellars Oldsville (Chehalem Mountains District)

Gran Moraine

Maysara Winery Asha and Declara (McMinnville District)

Penner Ash Estate Vineyard (Yamhill-Carlton)

Portlandia Hyland Vineyard

Ridgecrest Vineyard Estate (Ribbon Ridge District)

Sequitur (Ribbon Ridge District)

Sokol Blosser Estate Goosepen Block (Dundee Hills District)

Stoller Family Estate (Dundee Hills District)

16

No More Auditions

The day after *Sideways*'s slam-dunk premiere in Toronto, Thomas Haden Church had five movie offers. While making appearances during the awards season, the actor found himself delivering an acceptance speech at the Critics' Choice Movie Awards. Backstage he ran into director Sam Raimi, with whom he nearly worked a few years earlier on his supernatural thriller *The Gift*. After hugs and congratulations, Raimi asked to meet with him as soon as possible.

The two had lunch on the Sony lot. In his office, the director laid out his storyboards for the third of his *Spider-Man* comic-book movies for Columbia Pictures and walked Thomas through the whole movie in two hours.

"I want you to play Sandman," he told the actor.

Thomas eventually did, but not before he made several other movies, including James L. Brooks's *Spanglish*, along with TV shows. Thanks to *Sideways*, he was once more an in-demand actor.

"It changed everything for me," says Paul, the film's star. "I've never had to audition for another movie again. I don't think I would have done any of the stuff I did—even *John Adams*—if not for *Sideways*."

The seven-part HBO miniseries about the American Revolutionary War leader and US president, broadcast in 2008, won critical accolades for him again and another round of awards season wins. His career has taken off in film roles big and small, more stage work (especially *Hamlet* at his alma mater), television commercials, and a lead role in the Showtime series *Billions*.

Virginia Madsen experienced the same thing.

"Besides confidence, the movie gave me the opportunity that I never had before of being at the top of the list," she says. "It gave me offers rather than auditions. That's lasted. 'Oh, that's what she does,' people said. 'We want her.'"

For Sandra Oh, an actor who like Paul worked constantly before and after that movie, the *Sideways* awards season coincided more or less with the launch of *Grey's Anatomy*, Shonda Rhimes's television series on ABC focusing on the personal and professional lives of surgical interns and their supervisors. That show, in which she played Dr. Cristina Yang, ran for 220 episodes from 2005 to 2014. She has won multiple Golden Globes and later became the first actress of Asian descent to receive an Emmy nomination in the Best Actress category for her role in the British TV spy thriller–cum–black comedy created by Phoebe Waller-Bridge, *Killing Eve*.

"The combination of those two juggernauts [*Sideways, Grey's Anatomy*] did have an impact on my career," she allows, "but, no, I can't say because I was in *Sideways* solely that that led to anything else. That doesn't happen for everybody."

In 2005, at the height of the rush for independent filmmakers, Michael London was approached by investors from New York. His track record, which also included *Thirteen, House of Sand and Fog, The Family Stone*, and *The Illusionist*, was all modestly budgeted successful indies.

"They said we can probably raise money to make more movies like these," he recalls. "I said yes and was brought to New York by Goldman Sachs and we did like a road show where we created a sales documentary for a company, Groundswell Productions, where we would make a certain number of movies a year and would talk about budget projections. This was presented to potential investors and we raised the money for Groundswell. This never would have happened without *Sideways*."

And what about *Sideways*'s director?

"I'm the only one who didn't work for seven years," deadpans Alexander.

***

While he didn't direct a feature for seven years, those years did fill up with other things.

His writing partner defends him: "It's different for a writer-director than a director looking through scripts," says Jim Taylor. "We don't write that quickly either."

Among other things, his marriage to Sandra fell apart and divorce proceedings began. He moved out, eventually buying a house in the LA area and then an apartment in Omaha, did some prep for his segment of *Paris, je t'aime*, and traveled to several film festivals.

He headed the 2005 jury of En Certain Regard at the Cannes Film Festival, the official selection's largest sidebar of new films. The jury awards here are as closely scrutinized as the main In Competition selections, and many times the films are better.

At parties and events on the private beaches that stretch along the shore of the Mediterranean just off the Croisette, he appeared to be relaxed and enjoying himself, especially when in the company of Agnès Varda, the late and beloved French director and artist.

Meanwhile, as he prepped the Paris shoot, he and Jim began work on what would become a somewhat tortuous writing process to wrestle a science-fiction idea into a workable script, *Downsizing*. The story would take place on planet Earth, where a partial solution to overpopulation is the downsizing of humans to about five inches so they can live in relative luxury and splendor. The film would not get figured out, made, and released until 2017.

"*Downsizing* was a bear of a screenplay," he grumbles.

Paris, je t'aime, meant as a cinematic love letter to the City of Lights, was designed so twenty directors would have five minutes each for a short film, with each transition beginning with the last shot of the previous film and ending with the first shot of the following film and each transpiring in a different area of the city.

Fellow directors included Olivier Assayas, Gurinder Chadha, Ethan and Joel Coen, Alfonso Cuarón, Walter Salles, Tom Tykwer, Gus Van Sant, and Gérard Depardieu.

His sweetly melancholy segment, the last one in the film, *14e arrondissement*, finds a lonely middle-aged American woman (Margo Martindale) wandering the Paris of her dreams and thinking about her life (in terribly accented but studious French) and some of its disappointments, but realizing she's alive and well in Paris and it is the Paris of her dreams. She loves Paris and Paris now loves her.

Alexander was also dragooned into acting in another segment when an actor fell sick. As he, with very long dark hair in those days, walked by Wes Craven's office two days before his shoot, Craven suddenly pointed to his fellow director and called out, "What about him?"

Thus, Alexander played Oscar Wilde, risen briefly from the dead to lecture a groom about the need to demonstrate his love for his bride, in Craven's *Père-Lachaise* segment, taking place in that famous Parisian cemetery where Wilde and many other celebrated figures in the arts lie buried.

Fox Searchlight was so pleased in the aftermath of *Sideways* that the division offered Alexander, Jim, and their partner, producer Jim Burke, a deal to get first dibs on any project they developed. A production-development company, Ad Hominem Enterprises, was created, with Fox Searchlight paying for an office and staff in Santa Monica that Burke managed and a support person in New York for Jim.

"I thought it was worthwhile to have structure, mostly a support system for Alexander, but I didn't find it useful," says Jim. "We got rid of it on its tenth anniversary."

Nevertheless, they did wind up producing *The Savages*, a smart comedy written and directed by Jim's talented wife, Tamara Jenkins (*The Slums of Beverly Hills*), which starred Philip Seymour Hoffman and Laura Linney, and then the amusing character piece *King of California*, starring Michael Douglas and Evan Rachel Wood, by Alexander's old UCLA buddy, writer-director Michael Cahill. Later all three partners produced Miguel Arteta's *Cedar Rapids*, starring Ed Helms, John C. Reilly, Anne Heche, and Sigourney Weaver.

"Producing those films was a lot of work," Alexander says. "But it did need the imprimatur of Jim and me to tip them over into production."

And work continued on *Downsizing*. The two chipped away at the script for what became years. In the interim, "we took money jobs," says Alexander.

They worked on *I Now Pronounce You Chuck & Larry* (2007), a strained and predictable Adam Sandler comedy about two Brooklyn firefighters pretending to be a gay couple to get domestic partner benefits, and *Baby Mama* (2008), about an unlikely surrogate starring Tina Fey and Amy Poehler, but not much of their writing was used and the two received no writing credit.

Alexander took six months off to direct the pilot of the HBO series *Hung* (2009), about a high school basketball coach who takes advantage of his considerable genital endowment.

"The pilot of *Hung* came my way, where I had to rewrite a pilot," he says. "Jim helped me on that for a week or two."

With *Downsizing* showing no signs of yielding to their faithful ministrations, Alexander strongly felt a need to get back behind a motion picture camera.

Jim Burke had found a novel, *The Descendants*, which Alexander liked but not enough to derail his obsession with *Downsizing*. The outstanding British director Stephen Frears agreed to direct for Ad Hominem, so Jim Rash and Nat Faxon adapted Kaui Hart Hemmings's novel into a screenplay.

"When Frears backed out, I was finally so desperate to direct a feature I said, 'Let me take a crack at *The Descendants*,'" says Alexander. "I started writing the script on my own for about twelve weeks. Jim was busy and I didn't feel [the story] underneath the skin. I needed to spend alone time with the material to make it mine."

He spent most of his prep time in Hawaii getting to know that world, with novelist Kaui Hart Hemmings as his cultural guide. Shooting in early 2010 for a 2011 release, Alexander was back in top form with *The Descendants*.

***

Like *Sideways*, the movie takes place in a bucolic landscape, this time tropical, verdant Hawaii, wherein its characters encounter extreme physical and emotional distress despite the environmental splendor.

Matt's (George Clooney) disaffected wife has been in a boating accident that leaves her in a coma. Their ten-year-old daughter Scottie (Amara Miller) is acting out at school and elsewhere and seventeen-year-old Alex (Shailene Woodley), in boarding school on the Big Island, has rebelled against her entire family, especially her mother.

When Matt drags Alex back home to Oahu, she lets her clueless father have it: his wife—her mother—has been having an affair and was about to ask him for a divorce.

Matt desperately needs to reconnect with his estranged family, but his obsession over this adulterous affair engulfs him. It does, however, bring him and his elder daughter together in a kind of detective story as the two track down and confront the guilty party.

As with *About Schmidt*, the journey takes a viewer into the heart of a troubled man—his disintegrating marriage, a family adrift, a businessman cut off from his own life, and the needs of children with no guidance in their lives. As with *Sideways*, a flawed man is trying to figure himself out on the fly.

"Director Alexander Payne is a master of the human comedy, of the funny, moving, and messy details that define a fallible life," wrote Peter Travers in *Rolling Stone*. "Clooney has never exposed himself to the camera this openly, downplaying the star glamour and easy charm. Even the laughs come with a sting."

When Jim Gianopulos asked Alexander why he didn't cast Tom Hanks as Matt, the director answered, "He'd be wonderful but I've seen Tom Hanks cry. I've never seen George Clooney cry."

***

Two years later he had another new movie in cinemas. Before *The Descendants* went into production, *Election* producers Albert Berger and Ron Yerxa sent Bob Nelson's script *Nebraska*, which was set up at Paramount, to Alexander and asked him if he knew of any Midwestern director who might want to do it.

"Yeah, how about me?" he replied.

He rewrote the screenplay some, but Nelson retained sole credit. There had been regime changes on the Melrose lot, however, so the studio was barely aware they even owned the screenplay. Then when he told Paramount he wanted Phedon to shoot in black-and-white, the studio balked.

"I walked away from it," says Alexander. "I'll wait until another regime or two."

Not wanting to miss out on an Alexander Payne movie, the studio relented, but with $2 million less in the budget, he had to take the Directors Guild of America minimum for his services.

Nebraska presents yet another road trip. As with *About Schmidt*, he cast another aging refugee of New Hollywood circa mid-sixties to early eighties, Bruce Dern, as Woody, a burnt-out senior packing the heavy burdens of a life of despair. Dern plays a somewhat alcoholic and mentally addled elder who receives a $1 million Mega Sweepstake marketing brochure in the mail. It's a magazine subscription ploy, of course, yet he becomes convinced he has won the prize money. To humor him, his son David (Will Forte) accompanies him on a drive from his home in Billings, Montana, to Lincoln, Nebraska, to "collect" his prize.

The gorgeous and timeless black-and-white cinematography wonderfully captures the achingly austere Midwestern landscape in a manner reminiscent of what Robert Surtees's black-and-white cinematography did for hardscrabble Texas land in *The Last Picture Show*.

As you meet the family of monosyllabic, taciturn folk and encounter a very disturbing so-called former friend in Stacy Keach, you—and Woody's son—get a glimpse into what his old life was once like and possibly the torments and conflicts that created the man before you now.

June Squibb, so briefly seen in *About Schmidt* as Nicholson's wife, has a grand part as Woody's brutally honest wife, with the film's best lines. She is a snarky, foul-mouthed, formidable force.

Filled with deadpan humor and shot once again in his native Nebraska, the film is a bleak portrait of life in small towns yet done in a rueful, empathetic manner that fully displays the humanism that is a hallmark of all Alexander Payne movies.

***

Meanwhile, it was back to the coal mine that *Downsizing* had become. The film was originally intended to star Paul Giamatti. Alexander had remained in touch with the actor over the years.

"He is the greatest actor alive," Alexander asserts.

He invited Paul at one point to Omaha as a special guest at a nonprofit cinematheque called Film Streams for a retrospective of his films and then an interview with Alexander on stage in a large performing arts center seating 1,500 people. It sold out.

"I played a trick on Paul," admits Alexander. "I said people say, 'Oh, Paul Giamatti can read a phone book and make it work.' Then I reached under my chair and pulled out the Omaha phone book and handed it to him. 'Let's test this out. Would you please read something?'

"He read from the Omaha phone book *brilliantly* and brought the house down."

Alas, Paul's comic brilliance did not adorn *Downsizing* when it was finally made and released in 2017.

"Alexander told me about *Downsizing* when we were on *Sideways,*" says Paul. "Such a great, simple, satirical idea. He had a hard time to get it made; it was such an expensive movie. He had to get people who could get him the money. I really loved that script."

For a $65 million sci-fi movie-cum-social satire Paramount wanted a star, and Alexander thought Matt Damon was a good choice. He is a highly talented actor and very much a star, but Damon is not an ideal comic protagonist.

It's fair to say that Alexander and Jim never did beat the screenplay into submission. Beginning like a serious version of *Honey, I Shrunk the Kids*, the film struggles to evolve into social commentary. Matt Damon plays Paul Safranek, an occupational therapist who enthusiastically embraces the irreversible process while his wife (Kristen Wiig) goes along—until the last moment.

Some comedy evolves out of Paul finding himself on the other side of a permanent divide, then the movie takes off into an investigation of Leisureland, noisy neighbors, a wild party, two shady but fun-loving Eurotrash opportunists (played with oily smugness by Christoph Waltz and Udo Kier), an examination of Leisureland's class and racial divisions, and then a sharp turn into a "humanitarian" mission to Norway.

About halfway through, the movie gets hijacked by Hong Chau's Ngoc Lan Tran, a small-person Vietnamese activist whose physical disabilities Paul tries clumsily to minister to, only to fall in love with her. Chau's spitfire personality and indomitable will interject a dynamism into the film which until then it has lacked. You almost wish she had been the focus much earlier.

"The screenplay was such a huge idea, and putting twenty pounds of story sausage into a four-pound feature film with analog casing is tricky," admits Alexander. "I could do it better now as a limited series. As a movie, it winds up taking small hairpin turns and shaggy-dog curves."

Downsizing became the first Alexander Payne movie to lose money.

"I saw a documentary where Max von Sydow called George Stevens's *The Greatest Story Ever Told* [an epic Jesus tale which starred von Sydow], 'a fascinating failure.' I hope someday people will think of *Downsizing* as a fascinating failure."

17

The *Sideways* Effect

Not too long after *Sideways*'s box-office success and its rampage through the 2004/2005 awards season, journalists began using the phrase "the *Sideways* effect." According to California's Wine Institute, US supermarket sales of Pinot Noir jumped 18 percent in less than a year following the film's release. That, wine writers insisted, was due to the *Sideways* effect.

Sideways had become a cultural phenomenon. Its impact was local, as was the case for the Santa Barbara wine and hospitality community, regional in its influence on the California wine industry, and national as a hugely popular comedy as well as an unexpected guide to wine drinking in North America and, for that matter, the world.

Asked if he was surprised by the film's success, Alexander exclaims, "I still am! I thought it was a nice, slight little story. I had no idea it would become what it became and continues to be. I continue to be astonished—people seeing it twenty times, its cult status. Not just film people but wine people."

A movie about sports car racing will, of course, draw racing fans, but they were fans before the movie. A movie about deep-sea diving or big business shenanigans or politics or religion or science will naturally draw people interested in those subjects. But *Sideways* initially appealed to lovers of comedy or admirers of Alexander Payne movies or simply moviegoers who read reviews and wanted to see what all the fuss was about. Wine lovers may not even have known the film existed. Most filmgoers not only loved the comedy, they suddenly got a WSET (Wine & Spirit Education Trust) Level 1 crash course in wine.

No American movie has put the subject of wine—what wine culture is, how one appreciates the juice in tasting rooms or at dinner, how debates among oenophiles go, and how it pairs so nicely with romance—front and center. If American movies deal with alcohol or wine at all, it's as a problem, as in *The Days of Wine and Roses* or *The Lost Weekend* or as whimsy, as in *Arthur* or *Papa's Delicate Condition*. Or as in *A*

Walk in the Clouds, cowritten by Sonoma winery owner Robert Mark Kamen, where a vineyard might serve as a scenic backdrop for a postwar romance.

American cinema ignored wine as a part of American society until *Sideways*. Celebration scenes in movies usually called for bubbly, and that was it. In *Disclosure*, wine is a prop and a key plot point when Demi Moore tries to seduce Michael Douglas with a rare bottle of '91 Pahlmeyer Chardonnay that, in a later sexual harassment complaint, serves to back up Douglas's claim.

Even more recently, *Drops of God*, an eight-episode Apple TV+ drama, which streamed in 2023, set in the world of high-end wine about a wine-identifying challenge for an inheritance that includes a wine collection worth $150 million, focuses on the arcane, geeky, and snobbish aspects of wine, such as appellations, domaines, *négociants*, bouquet, tannins, vintages, and aeration. Romance? Forget it. The series is to wine what Netflix's *The Queen's Gambit* was to chess: the worlds of wine and chess are rendered exotic and unknowable to the casual viewer. No one is going to storm any wine region after getting through *Drops of God*.

What *Sideways* did, which its director may not have considered when enthusiastically grabbing an unpublished novel having to do with wine for a movie project, was introduce audiences to a subculture hitherto unexplored in American movies. By doing so, the movie made wine cool. It made the wine world look jolly and sexy. And it put a wine region, Santa Barbara County, on the map.

Suddenly, Santa Barbara's Harvest Festival was crowded with people who knew about only Napa and maybe Sonoma. Neophyte wine enthusiasts jammed local tasting rooms. *Sideways* maps appeared all over the county, directing people to the movie's many sites. Hotel space was at a premium on weekends. The dowdy Windmill Inn, where Miles and Jack stay, became the Sideways Inn and then got a makeover. Signs popped up in front of wineries and other places of business informing visitors that *Sideways* was shot there, even if that moment of glory happened in a split-screen panel.

"The funny thing to me is that everyone claimed to be in *Sideways*," remarks Fiddlehead's Kathy Joseph. "During the dinner when they pan the labels of the bottles in the racks, you're in *Sideways*. If they drove by your vineyard, you're in *Sideways*. I think everyone found a way to be a part of it."

Frank found himself encumbered with a fortunate problem, although it was still a problem. "That summer we were 40 percent busier than we had ever been," he says. "We had to be ready. We got new carpeting and new air conditioning and a new computer system. We took OpenTable. Turns out almost all our reservations were

Kathy Joseph at Fiddlestix Vineyard, Sta. Rita Hills, Santa Barbara County, California AUTHOR'S COLLECTION

The Windmill Inn forever transformed into the Sideways Inn AUTHOR'S COLLECTION

out-of-area-code numbers—213s, 714s [Southern California area codes]. All our locals were not calling ahead. We would block off tables so they could get in. We built our business on locals."

The same thing happened with Hitching Post wine, especially its Pinot Noirs.

"We had produced four thousand cases in 2003, the year the movie was shot, but we were not sure how we were going to sell it. The wine went from four thousand to fifteen thousand cases in two years. We've done as much as twenty thousand. We could have made a hundred thousand cases. The brand is potentially unlimited."

"All I can say is Frank Ostini has made a lot more money off of *Sideways* than I ever will," says Alexander.

Foxen's Dick Doré experienced the same rush: "I was seeing people I never saw before. People with tattoos and earrings in limos from Hollywood. A younger generation, twenty-one to thirty-five, instead of from fifty-five to seventy—a whole new generation of people. *Sideways* took the shine off the bottle and made it human. It made wine real to a lot of people. The tasting room went from two hundred to one thousand people a week. Sales skyrocketed."

Within a few years, the movie sparked an airline to add new routes to accommodate wine tourism. Alaska Air added these routes to their destination schedule, modeled after the very successful Wine Flies Free program ("take back a case of wine for no extra baggage fee") to and from Santa Rosa, Sonoma. The added routes were from San Diego, San Francisco, and Portland to San Luis Obispo. That Central Coast city is nearly equidistant to the wine tourism hubs of the Santa Ynez Valley and Paso Robles.

"We ruined that town," says the film's designer, Jane Stewart. "It was a sweet place, sleepy, quaint place, and now. . . . "

Actually, no town itself was "ruined," because Solvang was already a tourist trap best known for its Danish windmills, aebleskivers, and Viking-themed tchotchkes. Built long ago by Danish pioneers in the area, locals have always avoided it if possible. Los Olivos was already becoming a town of tasting rooms, and Buellton is scarcely a town at all, but rather a stretch of Highway 246 over the freeway.

Sam Marmorstein of Los Olivos Café, the scene of one of the movie's key sequences, says the *Sideways* "floodgates" startled everyone. "We were caught off guard—not enough staff. We had to staff up fast just to keep up with it. I had interviews on CNN and CBS *Sunday Morning* because of the movie. Chinese tourists knew about *Sideways*."

Another international event, however, brought further attention to the region and its wineries. The superstar American pop singer-songwriter Michael Jackson was criminally charged with molesting a thirteen-year-old boy at his Neverland Ranch estate in Los Olivos. *People v. Jackson* was a 2005 criminal trial held in Santa Barbara County Superior Court in Santa Maria.

"On off days, people from all over the world, representatives of newspapers and TV stations, with nothing to do came out to the wineries," relates Dick. "I was interviewed by New Zealand TV and Australian TV. We got national notoriety. We had a 30 to 40 percent increase in Pinot sales."

Inevitably there was pushback from vintners. They were pouring three times as much Pinot Noir at festivals and in tasting rooms. Everyone wanted to drink Pinot, but not everyone wanted to understand it or to buy it. It was the fad of that moment.

A tasting room assistant at Kalyra Winery, where Jack makes the dinner date with Stephanie, says that scene brings in customers weekly to this day. Many reenact and shoot with smartphone cameras the shot of Jack coming down the outdoor stairs with a case of wine.

"For a long time, I couldn't go to Santa Ynez because if I went to a wine tasting room it would turn into a three-hour photo session," confesses Virginia Madsen.

Winemaker Bryan Babcock shakes his head over these reenactors, who flooded tasting rooms just after the release, turning the valley into one big photo op.

"All these crazy people trying to relive the film," says Bryan, whose Babcock tasting room was inundated by hordes of fans. His theory, though, is that this explosion was predestined. "This region was ready to explode anyway. The movie just kicked it into high gear."

But fans reliving the movie, particularly certain parts of it, may have hurt the area.

Virginia sees the movie as having "changed the way people enjoy themselves. It opened up the world of wine and socializing differently. When people go up there by busloads, I think it hurts the feelings of winemakers when people are drinking just to get drunk. Go to a bar in the San Fernando Valley [in Los Angeles] and do that. To enjoy wine with friends, families, and strangers, it becomes a warm experience and should be shared."

"We didn't have the hospitality infrastructure, and I don't think we were quite ready to have a spotlight shown on us in quite that way," remarks Wes Hagen. "Some of our older winemakers realized the type of people coming up to taste our Pinot Noirs were not their type of customers, not lifelong customers. They were riding the cusp of a trend.

"I like to say 25 percent of the trendies are still our friends and customers and wine club members. So, in the end, it was a positive, growing experience for Santa Barbara County. But at the time it was a little dirty and not quite the way we wanted to be known for the quality of our wine."

***

There is little question of *Sideways*'s impact on Pinot Noir. When Miles deprecatingly says of Jack, "Yesterday he didn't know Pinot Noir from film noir," he might as well have been speaking about many of the movie's viewers. Pinot Noir wasn't in many wine conversations, other than about Burgundy or among wine geeks such as those hanging out at Epicurus. California's heavy red hitters then were Cab, Merlot, and Bordeaux-style blends, and in a minority report, Zin had its vociferous fan base. Wine experts all picked Syrah to be the next breakout grape.

Never happened. Instead, *Sideways* happened.

"People could actually pronounce Pinot Noir after that movie," says Oregon Pinot winemaker Harry Peterson-Nedry (Ridgecrest Vineyard). This is not a casual or cynical remark but goes to the heart of how Americans drink wine. They order what they can pronounce, which is one reason why Merlot was so popular and Gewurztraminer, a wonderfully fruity wine with expansive lychee aromas and flavors, has never caught on.

The A. C. Nielsen Company compiles data that tracks "supermarket sales" of all products identified with universal barcodes. For the thirteen weeks ending in early June of 2005, compared with the same period a year earlier, Nielsen saw that supermarket sales of Pinot had risen 83 percent in dollar terms and 77 percent by volume—one of the largest year-on-year increases in the history of wine sales data. These double-digit increases persisted, year on year, from 2005 to 2006 and from 2006 to 2007, although the slope of the curve flattened.

Given that high-end Pinot Noirs are often not sold through supermarkets, the Nielsen data, if anything, understates the real increase of Pinot, especially in dollar terms.

Wine & Spirits magazine produces an annual poll that gathers data from three hundred restaurants in major markets throughout the country and specifically documents sales in the last calendar quarter of each year. The magazine found that Pinot, as a percentage of top-selling wines overall, increased from 10 to more than 13 percent from 2004 to 2005—a more than 30 percent increase—and by another 2 percent from 2005 to 2006, with almost 90 percent of this rise associated with American Pinots and not Burgundy or other imports.

"Pinot Noir production in California has increased roughly 170 percent since *Sideways* was released," wine industry analyst Gabriel Froymovich of Vineyard Financial Associates told NPR in 2017, noting that total wine grape production increased 7 to 8 percent during the same time.

Although there is no reliable data on the number of wineries growing the grape, Pinot Noir historian John Winthrop Hager estimates that "the number of commercially visible producers grew from between 400 and 450 in 2003 to between 950 and 1,000 by the end of 2006."

In the two decades since the film's release, prominent wineries already in the Pinot Noir business have worked hard to further define the state's prime AVAs for Pinot, as well as lobbying for new ones such as the SLO Coast. A new generation of dedicated vintners has arrived, increasing vineyard-designated bottlings to highlight California's varying terroirs. A greater understanding of clones has fueled new plantings to complement the older vines, which deliver smaller yields but oh, such delicate, silky, luscious wines.

It wasn't just in Santa Barbara. In Monterey County's Santa Lucia Highlands, Joe Alarid, proprietor of the Tondré vineyard, whose 104 acres of grapes—Chardonnay, Syrah, and Riesling but the majority in Pinot—are in high demand, started getting frantic calls for more of his Pinot. "The demand was unbelievable. After the movie, people were calling me up: 'Hey, got any Pinot?'"

The *Sideways* effect hit David Adelsheim in Oregon's Willamette Valley in both the short and long terms. His current release sold out so quickly that he, a major Pinot producer, had no Pinot to sell for six months.

But the "real story about *Sideways*," he says, occurred during two sales visits four years apart. He recollects a Hollywood liquor store in 2002 turning down his Pinot not because it wasn't good but because the store's customers preferred Cabernet and Chardonnay. "Not much you can say to that," he muses. Then during a visit to a New York restaurant in 2006 he poured a Pinot Gris and was again turned down because "our customers are very picky. They only like Cabernet, Chardonnay, and Pinot Noir," the wine buyer insisted.

"That's what *Sideways* did. Not just in that restaurant but around the US," he says. "And today selling Willamette Valley Pinot Noir outside the US is easier than it's ever been, in part because Burgundy has become more expensive."

***

But when *Sideways* aficionados and reenactors reached for a bottle of Pinot in the wine store, which one did they grab? It mattered. And, no doubt, led to some disappointment. Many American palates, used to Cab and Zin and even Syrah, find many Pinots too "light." If it's varietally correct and if no one snuck any Merlot or Syrah into it to darken the color, the wine in those days might have appeared paler than the casual wine drinker was used to, and then, sipping it, he or she might miss the power of a Cab.

As far as Wes is concerned, the *Sideways* crowd that went from, say, Sutter Home White Zinfandel to a fine Santa Barbara Pinot may have traveled outside their comfort zone.

"Before *Sideways* Pinot Noir required wine drinkers to reach the end game of appreciation of wine," he says. "You start with sweet wine, then dry wine, have a Cab phase, have Rhône phase, and when you get everything else figured out—oh my God, there's this thing called Pinot Noir that will break my fuckin' heart. So you go through the process of a heartbreak but say, damn it, I *will* drink Pinot Noir.

"The *Sideways* effect did for Santa Barbara Pinot Noir what the Soviet Union did for socialism: We went from feudalism to postmodernism with no work in between."

Wes's understanding of how Pinot translates to casual drinkers may be more astute than his political science, yet Pinot may well have been a bridge too far for some.

Many, however, did succumb. Plus, the smell, taste, connoisseurship, and the movie itself told them this was fabulous—and they truly agreed.

"Pinot Noir is subject to real intellectual pursuit within the wine industry," Master of Wine Clare Tooley points out. "You have extraordinary proponents of the grape—winemakers, wine writers, merchants who made Pinot Noir their holy grail. Over decades if not longer it's had that sense of importance. So there's a certain validation from professionals that this is a great variety to be chameleon-like, to take on a sense of place, and yet retain its innate characteristics.

"On top of that, marketing influences an audience—movies, books, common parlance—and when the taste does have a certain aroma, softness, lightness, and also it's quite jammy and rich, the flavor and aromatic profile are very appealing regardless of whether you know about it or not. So the combination of the intellectual and mass marketing created a rather perfect storm."

"That film changed entirely the trajectory of our industry in the United States for quite a while," says veteran winemaker Adam LaZarre. "There has never been such a jarring change of direction. My belief is that everything is on a ten-year cycle except Cabernet and Chardonnay, but everything else kind of ebbs and flows. Go back the last thirty years you can see where Zinfandel peaked and then ten years later Sauvignon Blanc. But this just changed everything in about a month and a half. It was that jarring."

***

Then there is this dialogue exchange:

> Jack: If they want to drink Merlot, we're drinking Merlot.
> Miles: No, if anyone orders Merlot, I'm leaving. I am NOT drinking any fucking Merlot!

The film opened in late October 2004. As early as Christmas, sales of Pinot Noir soared while sales of Merlot dropped. It is not unprecedented for movies to affect, for a brief while, other areas in popular culture.

The old story goes that when Clark Gable unbuttoned his shirt in the 1934 screwball comedy *It Happened One Night* to reveal that he—The King, Hollywood's number-one box-office attraction—wore no undershirt, a similar thing happened. Whether this scene was the cause or not, men's undershirt sales went into decline. Some think undershirts were heading that way anyway.

Game of Thrones, HBO's hugely popular series adaptation of George R. R. Martin's fantasy novels, in which characters enjoy ceremonial horns of mead, first aired in April 2011, when the number of commercial meaderies in the United States was less than 200. Today there are at least 480, according to *Smithsonian* magazine.

The difference with the *Sideways* effect on Merlot, however, has been its staying power.

"To this day, *Sideways* still has a negative impact on Merlot," says Wes.

"The stigma attached is still strong," agrees Clare, who works as VP of guest experience for the Boisset Collection in Napa.

How did this happen?

Merlot is one of Bordeaux's most celebrated varieties. Merlot ripens early with large grapes. It's thin-skinned and known for its full-bodied yet soft fruit. The tannins are easy, and there is often a soft finish that makes it an ideal blending grape. In Bordeaux's Right Bank regions of St. Émilion and Pomerol, Merlot reigns supreme.

"Merlot was born for the American palate—softer than Cab, not as aggressive so it won't slap you in the face with those bell pepper aromas," says Wes. "It hits the American palate with what it needs—richness, depth, saltiness, fruit, and a sweetness to it."

To Alexander this aspect of the *Sideways* effect has been a "huge surprise. There was some version [of that line] in the book we turned into a pretty funny joke. Never in a million years did I think that *one joke* would tumble the Merlot empire."

And yet there are some Merlot producers and admirers who are happy about that joke. It helped weed out some very bad Merlot.

So we put it to Rex, whose novel continually disparages Merlot, even if that line—which Alexander says will be "quoted to us the rest of our lives"—was cut by an editor: does he hate Merlot?

"No, of course not," he snorts. Rather than a *Sideways* effect, he puts Merlot's demise down to, in a roundabout way, a *60 Minutes* effect.

One Sunday evening in 1991, that CBS news program highlighted the French paradox, a theory positing that wine contributed positively to the health and longevity of the French population despite their generally high-fat diet. Morley Shafer introduced 21.8 million Americans to the puzzling reality that the French, consuming far more saturated fats, had better health than Americans. The term was coined by French nutrition and cardiology researcher Serge Renaud, who claimed that regular, moderate red wine consumption—particularly with meals—was behind the healthier French hearts. US red wine sales jumped a whopping 39 percent in 1992.

"The study was totally flawed because the French ate better foods than we did," says Rex. "But Americans searching for the fountain of youth rush to liquor stores: 'I want red wine!' They get Cabernet Sauvignon, which is tannic and astringent. They rush back. 'Give me something else because I just need red wine and want to live forever!' Thus, Merlot—it's fruitier, more approachable, easier on the palate—and so was born the vitamin E of the nineties.

"Merlot gets overcropped, planted everywhere. By the time I'm going to Epicurus with wine snobs, Merlot is tantamount to being a wine philistine. If you order Merlot, you don't know anything about wine. Merlot earned that reputation."

Before *Sideways* hit, at a restaurant or bar with wines by the glass, it was all too easy to default to Merlot for a red wine and Chardonnay for white. Chardonnay can be a little bit sweet, like Merlot. So those became the two wines ordered by people who didn't know or care much about wine.

Most winemakers today acknowledge that too much Merlot was being planted back then, and much was green and boring. It was being planted in soils that it should never have been planted in. Yes, it's easy to grow, but that doesn't mean it should be planted everywhere. So some bad grapes went into wine that was mediocre, thin, and too sweet, but an affordable, fruit-forward "friendly" wine.

"I hope Merlot never gets back up to the popularity it had in the nineties," says Adam. "I could walk into a friend's winery and ask him what he's bottling and he says, 'Oh, I'm bottling Merlot,' and I say, 'This is garbage.' He says, 'It doesn't matter. It says Merlot on the bottle.' He's going to sell it out.

"The movie did a service to the grape. As sales plummeted to almost nonexistent levels, what was left behind was quality. Now I see the wine buyers, the people who make decisions for restaurants and chains and retail places, are all young kids, all twenty-five-to-twenty-six-year-olds with WSET Level 3 [certification] or master somms. Many of them never saw *Sideways*. You mention the movie and they say, 'My dad saw that and told me about it.' So, for a lot of them, Merlot is a fairly new and exciting grape."

Clare defends Merlot as a "supremely elegant, noble grape variety" but admits that it attained popularity with consumers because it was easily pronounced, and popularity with winegrowers because it "ripens early, is pest resistant, easy to grow and plant, but you can overcrop it and get high yields. It suffered from that. The industry went down that road and made a lot of Merlot. It was a race to the bottom."

Rex insists the line may have been misunderstood.

"What's happening in the movie?" he asks. "Jack is about to go in [the restaurant] and cheat on his fiancée, and Miles is going to get blamed for it. So Jack is giving him all this advice on how to act so this deal goes down, right? He's worried about Miles because he's socially awkward.

"'If [the women] want to drink Merlot, we're drinking Merlot,' he says and then Miles goes into this rant: 'No, I'm leaving.' It's his way of saying, 'I'm a loose cannon and your night may not go well. So stop with your fucking advice.' So, he's really try-ing to stop the train wreck of Jack in that moment."

Misunderstood or not, the line still changed the American wine industry. Adam even admits that "a lot of bad Merlot plantings were ripped up and replaced with bad Pinot. Wineries signed fifteen-to-twenty-year contracts [with those growers] with unsustainable pricing. After the movie, people had no problem paying twenty or thirty dollars for a bad bottle of Pinot Noir because it said Pinot Noir. Now big wineries are stuck with these contracts."

This means that Pinot, like Merlot, could become a victim of its own success. Some feel—we'll get to this—it already has.

It is estimated California growers pulled out more than 10,000 acres of Merlot after the movie. At its peak in California, Merlot accounted for nearly 60,000 acres of vines. The Wine Institute's 2022 California grape acreage report shows 33,808 acres of Merlot planted.

(As a frame of reference, Pinot Noir plantings in California are 47,153 acres. Com-pare that to Burgundy's 26,277 acres devoted to Pinot. The largest plantings in Cali-fornia are in Sonoma, followed closely by Monterey. Santa Barbara is a distant third.)

Some wineries, especially in Napa Valley, stuck firmly with Merlot, especially those who had earned reputations for making fine Merlots. One of those was Trefethen Family Vineyards. For three generations the family has been one of the pioneers in the Napa wine community. Cofounders Janet and John Trefethen traveled the world promoting Napa wines as well as their own. Merlot was always a personal favorite.

Merlot was planted in the early 1960s in Napa's Oak Knoll appellation but wasn't produced as a varietal wine until 1993, explains second-generation vigneron Hailey Trefethen.

"It took us a while to dial that in for it to stand on its own," she notes of the twenty-four-acre Merlot planting that has been fine-tuned over the years.

Not long after the movie came out, the family had a billboard erected on the southwestern corner of their Oak Knoll vineyard, which directly faces Napa's iconic Highway 29. The sign read, "Clearly, Miles Never Had a Napa Valley Merlot."

***

Napa's Duckhorn Vineyards built its reputation and success on Merlot. Dan and Margaret Duckhorn returned from a trip to the St. Émilion and Pomerol regions in Bordeaux, having fallen in love with Merlot as a variety. They were drawn to its versatility and food friendliness. Back in Napa, they focused on Merlot as a varietal wine, which no one was doing in the seventies. Harvested in 1978 and released in 1979, eight hundred cases of Three Palms Vineyard Merlot hit the market as a stand-alone luxury varietal at $12.50 a bottle, a sky-high price then.

"Dan would drive up and down the coast of California selling it out of the trunk of his car and developed a following," says Renée Ary, the fourth winemaker for Duckhorn in over forty-five years of winemaking.

"We got a little dip [in sales] after *Sideways*, but it kicked back up. We were always committed to the varietal, so we weren't about to throw that out the window because of a movie. For us, it was a positive change. It helped weed out some of the less serious or mediocre Merlot producers. And it freed up quite a few vineyards we hadn't previously worked with, and we were able to take advantage of that."

Duckhorn Vineyards deserves kudos for ingeniously resurrecting the much-maligned variety when it launched #MerlotMe in 2013, a movement supported by several Napa producers. Thanks to the Duckhorn leadership, a global band of vintners growing Merlot in great areas turned to social media to help rescue the grape's reputation. The month of October brings its annual #MerlotMe celebration, with tastings and events planned in many countries and states, culminating on November 7 as National Merlot Day.

Although the Duckhorn Vineyards is no longer funding the #MerlotMe project, the movement continues its global outreach. The #MerlotMe stats, according to Keyhole tracking data, show a total of 268,729,298 social impressions from 2014 to 2022.

"They've done a great job—their Three Palms Merlot is a fabulous wine," says Chris Carpenter, winemaker for La Jota Vineyard Co. and Mt. Brave in Napa, in a nod to the Duckhorn Vineyards. "They weathered the *Sideways* debacle well."

Chris is bullish on California Merlots. "There's no reason why France and Italy should own the category," he insists, naming Right Bank Bordeaux's Petrus and Tuscany's Masseto as perhaps the most renowned Merlots on the planet. "We have in Napa these great areas that have incredible Merlots. I'm proud of our Merlots and think they can stand with the best of 'em."

Chris himself crafts powerful, plush, structured Merlots from the mountains surrounding Napa Valley. Their shallow soils give structure to the wines, while the sunlight helps to moderate the herbal component.

At Markham, the fourth Merlot producer in Napa Valley, winemaker Kimberlee Nicholls passionately says, "We pride ourselves on great Merlot, knowing how to grow it and how to craft it. So as a Merlot house, you have to work harder to make fans of Merlot."

Still, if people do purchase a good bottle of Merlot, more than likely it has Napa on the label and is usually from a recognizable winery. As Clare puts it, "Duckhorn does a good job of promoting Merlot. But do people buy it because it's Merlot or because it's Duckhorn? The significance of brand here [in the United States] you can't underestimate."

In Paso Robles, Tom Myers, veteran winemaker at Castoro Cellars, weighs in. Until the 1990s, Merlot was not known as a varietal wine. "Cab was king," he says. "Then [Merlot] took off; it was softer, marketable, and had a broader appeal." At Castoro, Merlot planting didn't begin till the late 1990s. "Then we were hit by *Sideways* and took a back seat for a decade, but we didn't pull it out; we found a home for it."

***

Shortly after the movie came out, Alexander received a package containing a magnum of Trefethen Merlot and a note that read, "Dear Alexander Payne, I'll bet Miles and Jack never had our Merlot. Warm regards, Janet Trefethen."

Alexander chuckles over this. "The key wine in the film is the '61 Cheval Blanc, which is mostly Merlot," he says. On the Internet, the average price for this particular bottle is $4,755. One site demands $9,423.

So Miles loves wine and doesn't hate Merlot. He just hates the lousy version of it habitually served in bars and restaurants. He declares at one juncture, "I like all varietals." Throughout their bachelor-party imbibing, the men slurp their way through a wide variety of wines—rosé at the Stanford tasting room, a Cabernet Franc launches Jack's wooing of Stephanie at Kalyra Winery, Sauvignon Blanc gets name-checked at the top of the Los Olivos dinner sequence, and Miles's soliloquy to Pinot happens over glasses of Andrew Murray Syrah.

Merlots from Napa Valley, California AUTHOR'S COLLECTION

What *Sideways* did was show people there was a whole world of wine out there, and it gave those wine curious permission to branch out. And perhaps, after listening to Miles disparage Merlot and Cab Franc, also permission to scorn wines they didn't like.

Most importantly, *Sideways* showed people a new way to talk and think about wine.

"What I learned from that [*Sideways*] experience was the freedom to engage with wine in a very curious way," says Sandra Oh. "If I'm in a restaurant with a somm, I engage with the somm: 'I had a really bad day. Can I have something that will lift my spirits?' I know I have the freedom to talk about it. When you talk about it and engage with a somm to really try to figure out what I like. I think that's interesting."

***

Okay, so here's the urban myth about the *Sideways* effect: moments after Paul utters his infamous denunciation of Merlot, sales of that wine in the United States plummet. And not long after moviegoers leave the theater, nearly everyone rushes out to buy Pinot Noir, because that wine was so highly and constantly praised throughout the movie.

For instance, George Schofield in the April 2008 issue of *Wine Business Monthly* refers to the effects of the movie on Merlot as the "debacle following the release of the *Sideways* motion picture."

More recently, James Molesworth, *Wine Spectator*'s lead taster on California Pinot Noir, writes in the September 30, 2022, issue, "Ever since consumer passion was ignited for California Pinot Noir by the movie *Sideways* nearly two decades ago, the category has seen a rapid upward trajectory."

It is a fact that plantings of Pinot have been increasing for a couple of decades in California. But Merlot has endured its roller-coaster ride and is gaining respect. It's the fourth-leading red wine purchased by Americans after Cabernet Sauvignon, red blends, and Pinot Noir, according to Nielsen data.

So if you eliminate "red blends," which might contain a lot of Merlot by the way, it's the number-three variety behind Cab and Pinot for varietal wine purchases. Number three, mind you. Where's the debacle?

While sales, plantings, and price points can be tracked by industry data collectors, much of the effect of *Sideways* on wine has become folklore supported by scant anecdotal evidence. Steven S. Cuellar, professor of economics and wine business at Sonoma State University, decided to test this urban myth by examining the actual effect of the movie on US wine consumption.

He published the results in the January 2008 issue of the *Journal of Wine Economics*. Such papers do not make for easy reading, as they swim in data from retail chains, empirical analysis, and equations. Also, as Dr. Cuellar tells us, a couple of interesting graphs didn't make the paper "because editors go through papers and turn them down."

The wines he studied against Pinot Noir are Cabernet Sauvignon, Chardonnay, Merlot, Pinot Grigio, Riesling, and Shiraz/Syrah.

He does make the not-insignificant point that at the time of *Sideways*'s release, Pinot Noir was on the upswing, while Merlot and other varieties were winding down. But he's mostly agnostic about The *Sideways* effect.

"It's really hard to see it having *that* big an effect," he says. "It could be that everybody who drinks wine saw the movie. You can see the growth rate of Pinot Noir actually slows down a little bit after the movie. It's something of a mystery."

Let's repeat that: Pinot "slows down." You would think the growth of Pinot would increase as opposed to decrease, as it does in the unpublished slide he shows us. Yes, it's going up, but Pinot was going up much faster *before* the movie came out.

"If Pinot Noir is pulling from all those other wines, you could expect the growth would increase rather than decrease after the movie. If it's pulling from other wines, why isn't it growing faster instead of slower? I really don't have any explanation for it."

No wine's growth rate increased after the movie. Curiously, of all wines he studied from 1999 to 2008, only Pinot Grigio and Riesling remained relatively unaffected. The growth rate for everything else went down.

"It doesn't look like anyone is benefiting from the decrease in Merlot consumption—at least among these seven varietals. It's an interesting cultural phenomenon."

Then why did growers rip out Merlot and plant more Pinot?

"They're responding to something that didn't happen," says the professor. "They're responding to the idea that people are buying less Merlot because the movie came out. But they were buying less Chardonnay and Syrah and even slowing down the growth rate of Pinot Noir. If someone yells 'fire' and there's no fire, people are still going to run out. They're responding to a false narrative."

Pinot Noir from Anderson Valley, Mendocino County

Northern California's Mendocino County evokes an unhurried pace amid meandering country roads and small towns, with vineyards and wineries ensconced in remote corners.

However, it's a fifteen-mile stretch of Anderson Valley that ranks among California's top Pinot growing regions. Commercial wine growing began in the 1940s but didn't kick in till the 1960s and 1970s, with plantings of Pinot Noir and Chardonnay. In 1982, when the region received its AVA status, there were six wineries with some five hundred acres under vine; now it's grown to more than two thousand acres with fifty-four wine producers.

Our suggested Pinots of Anderson Valley:

Domaine Anderson Dach and Walraven Vineyards
Handley Cellars RSM Vineyard
Goldeneye Confluence Vineyard
Husch Vineyards One Barrel
Maggy Hawk
Navarro Vineyards Méthode à l'Ancienne

Pinot Noir from Anderson Valley, Mendocino County, California © ZW IMAGES

18

Vertical

The year following *Sideways*'s debut, the authors ran into Rex Pickett at the Paso Robles Zin Festival. He was the guest of honor at that wine festival, where he would speak and autograph copies of his novel at Justin Winery, a devoted Bordeaux house, of all things. The Pinot lover even feigned indignation at having to stay at the winery's guesthouse, the Bordeaux Room.

Since the movie's release, Rex has been a guest at wine fests, has been asked to speak about his adventures in wine at other events, created his own Pinot brand, and has written about wine for various outlets. But the bitch goddess of fame and fortune has eluded him.

When *Sideways*'s production finished, his novel was still unpublished. Under pressure, he says, from Michael London, he resubmitted to publishing houses the novel that was now at least the basis for a film getting a major release. He got one offer—from St. Martin's Press for $5,000. He took it.

"That was the stupidest yes I ever said in my life," he laments.

He points to Amanda Brown's *Legally Blonde*, a self-published novel that sold for $1.5 million after that movie came out in 2001. Which is not to say the two situations are analogous, but he is undoubtedly right in conjecturing he would have gotten much more than $5,000 had he waited for the film's glorious reception. But then again, who knew?

But Rex was left with a check for the sale of the movie rights and—a brand. *Sideways* became a series of novels and a play that has been performed globally, and a musical version of that play is in the works. A concert of the songs already written for the musical was staged at Buena Vista Winery in Sonoma Valley in September 2021.

In 2011 he published *Vertical*, his first sequel to *Sideways*. It starts, as sequels are wont to do, following the further adventures of Miles and Jack in debauchery, womanizing, and wine guzzling. This time the boys drive through the scenes of their earlier notoriety in the vineyards of Santa Barbara County on their way to the International Pinot Noir Celebration in McMinnville, Oregon.

In *Vertical*, Rex exploits three parallel worlds: the novel, the movie, and his own life. Miles, Rex's alter ego, is forever marveling at his sudden celebrity. Women want to bed him and cases of hard-to-find, expensive Willamette Valley Pinots pile up in his hotel rooms, sent by wineries hoping for product placement in the sequel to *Shameless*, the novel the fictional Miles wrote.

"Sometimes reality and fiction get so intertwined in my wine-addled memory I couldn't tell if I was coming or going. Lying or not lying," Miles muses.

In this way, Rex updates his readers on his life since *Sideways* and announces a new theme: how does a wine-addled novelist untangle the deeply entwined threads of these worlds so he can go from sideways (drunk) to vertical (sober and functioning again creatively)?

This was the struggle Rex was going through himself as he was penning the novel. In 2011 he gave up drinking. Not tasting wine, mind you, but drinking it.

"I just got to the point where it got to be too much and people were taking advantage of me," he says. "I gave it up, but I sip and spit with passion, and on an almost daily basis.

"Alcohol, like other drugs, can be liberating for many artists. I never wrote drunk, or even with a single drink in me, but I miss the anesthetizing effects of it in the evening when, like most writers, we want desperately to turn our minds off. But when it begins interfering with your life, then it becomes, at least it did for me, an imperative to abstain and take back control of my life, and the intellectual property I created, and that's why I quit. But I love wine as much as I ever have.

"I love talking about viticulture. I'm as excited about it as I ever was, even if I don't drink. I often say I don't like being around sober people and I don't like being around drunks."

***

Most sequels, certainly in movies but also to a degree in novels, are remakes, that is, characters, themes, and situations get repeated so as not to annoy the fan base. To his credit, *Vertical* is not that kind of a sequel. Rex used his old screenplay, *The Road Back*, based in large measure on his experiences with his wheelchair-bound mother, as the basis for a decidedly different story.

The novel's road trip puts four troubled human beings plus a dog into Miles's rental ramp van heading for the Pinot festival. Along with Miles, rolling in dough, and Jack, now divorced and so permanently pickled he can no longer get directing work, are Miles's stroke-addled mother, Phyllis; her pot-smoking Filipina nurse, Joy; and his mom's Yorkie, Snapper. Miles schemes to snatch his mom from the assisted-living facility she hates and take her, via Oregon, to her sister's home in Wisconsin.

Rex doubles down on the pain quotient in his new story. Jack's overdose of Viagra while drunk in the sack with a hot babe results in a terrifying case of priapism. This results in a trip to the ER that would make a horror film buff blanch. There is also a tooth extraction, a badly injured dog, Phyllis's bout with rampaging diarrhea, and a dunking contest that deposits Miles in a vat of Merlot, where he is attacked by sex-crazed female fans of *Shameless*.

So the new novel is not a buddy story but rather a mother-and-son story. In the last chapters, as the journey heads from Oregon to Wisconsin, you sense the writer, clearheaded and determined, narrowing his focus on Miles and his mother. He insists they better understand each other and resolve a lifetime of distrust and hurt feelings. These passages contain the most powerful writing in the book.

***

The following year the authors attend *Sideways the Play* at the Ruskin Group Theatre, a small theater and acting school tucked away in one of the old buildings that line Airport Avenue at the Santa Monica Airport. An hour before every curtain there is a Pinot tasting featuring wines from small, artisanal producers, as Rex is determined to introduce his *Sideways* fan base to unfamiliar labels. On this night, we sip two luscious and spicy 2009 Mendocino Pinots from Waits-Mast Family Cellars out of San Francisco.

Rex is in a buoyant mood, as the play has sold out every night so far and the run has been extended. He tells us wine districts in several countries have approached him about a third *Sideways* novel set in their regions. They would fund his research, but he would have control over the content.

Patrons are encouraged to take plastic cups of Pinot into the theater, and the tasting continues at intermission.

The fifty-two-seat theater has an extremely tight space, yet director Amelia Mulkey sees to it that the play flows smoothly throughout the Santa Ynez Valley's various tasting rooms, bars, motels, and restaurants. With the audience seated in an L and a raised platform at the far end, the cramped stage uses a host of props, one ubiquitous bar counter, two Murphy beds, winery signposts, and more wine bottles and glasses than you can count to indicate the story's many locales. The well-coordinated set changes and repositioning of props by the ensemble cast, who also play multiple roles, are nearly as amusing as the play itself.

One cast member does an ingenious job of watering down fruit juices and teas and adding coloring to simulate the various white, red, and rosé wines being poured during the show.

The play ran for nearly one hundred performances.

Sideways: Chile arrived in 2015. While Rex did have, as promised, control over the content, he says he felt somewhat handcuffed by his agreement and decided against the caustic comments Miles might have made about the nostalgia for the Pinochet dictatorship he encountered among right-wing elements in Chilean society. It's also a novel, once it gets out of Santa Barbara County and down to South America, without much Pinot. *Sideways: New Zealand*, published in January 2024, finds Miles reunited with Jack on a "book tour from hell" in a camper from the south of the South Island to the north of the North Island with his Māori publicist in tow.

Sideways the Play has taken on a life of its own. It was staged by three-time Tony Award–winner Des McAnuff (*Jersey Boys*) in a run at the prestigious La Jolla Playhouse on the UCSD campus in 2013; its European premiere occurred at London's St. James Theatre in 2016; then Rex flew to Riga, Latvia, in October 2023 for the opening night of his play at the New Riga Theatre. The play's Latvian adapter and director, Alvis Hermanis, retitled the show *Pelicans and Grapes*. There's a scene in act 2 where Miles is pulled out of the ocean by Jack and launches into an introspective monologue in which he refers to the simple life of pelicans.

Despite not understanding a word of Latvian, Rex felt this was the "ne plus ultra version of my play" where a minimalist set, simple transitions, and the placement of actors heightened the emotional and comedic effect. In Rex's opinion, Hermanis put all the attention on his actors and the text.

From Latvia, Rex journeyed to Spain to hook up with the cast of his play there, where a ninety-minute somewhat sitcom-y version of the show had been on a national tour for two years. Meanwhile, a recording of *Sideways the Musical* songs was released on all streamers on August 18, 2023.

"The play was also staged in Tokyo, but I had nothing to do with that production," he adds.

Speaking of Japan, a nation that loves its wine and especially Pinot Noir, the film itself suffered a 2009 Japanese remake set in Napa Valley. Alexander says he did try to watch it: "I could take about ten minutes of it."

19

We Need Another Movie

Thanks to *Sideways*, wine became hip and romantic but, most importantly, *approachable*. There was true brilliance in having as Miles's partner in crime a fellow like Jack, a guy who doesn't know to take gum from his mouth to taste wine. As Miles educates his clueless friend about how to eyeball, sip, and savor wine, he educates the audience, too; Miles introduces the lingo and procedures of the wine tasting room to many casual wine drinkers, or even those who had never thought about wine before. Miles's colloquies with Jack about wine sweep away the snootiness and pretension many associate, wrongly, with wine drinking.

So what's the winescape like today? Where does Pinot Noir stand twenty years after *Sideways*? And what about Merlot?

"We've gone through a period of oversupply. The response [to the film] was to plant more," says vintner Bryan Babcock. "You'd see letters from brokers indicating there's one hundred tons [of] Pinot Noir for sale here and four hundred tons for sale there. It definitely became industrialized."

"We've saturated the market [in Pinot] in my opinion," says Joe Alarid of Tondré.

Acreage that was planted in Pinot due to the perceived notion that the sky was the limit for the variety thanks to the movie, meaning in 2005 or so, would have come into production around 2008 or 2009—just in time for the economic downturn. That in itself would account for some of those letters from brokers.

Merlot, meanwhile, has recovered remarkably and with much better Merlot, although it is not as ubiquitous as before.

"We've experienced an upward trend in the last fifteen years," says Renée Ary, Duckhorn's winemaker. "When you go to a restaurant, the Merlot category is small. So there are not that many players but we continue to see positive growth. Merlot is on an upward trend."

"Merlot is one of the great varieties if planted in the right climates where it expresses itself," insists Eyrie's Jason Lett. "Breaking things down into categories of good varieties/bad varieties—how silly."

There was a time, he notes, when most world wine regions never put varietal names on wine labels. A Bordeaux bottle lists the chateau and AOC but never the grapes; the same for Burgundy. Everyone understands what the blend (in Bordeaux) or the grape (in Burgundy) is. California initially followed suit, often expropriating foreign wine regions against their bitter objections for labels such as Hearty Mountain Burgundy for reds or Chablis for whites.

Then in the sixties, Robert Mondavi, among others, started to put varietal labels on wines. Now Europeans are doing it, at least when exporting to the New World.

"I sometimes wonder if we didn't put ourselves in a box by labeling," Jason muses. "It has led us to where one variety becomes ascendant over another and people are disregarding the region it came from as being a contributor to ways the wine tastes. Miles never said, 'I hate Bordeaux.' So much of this has to do with region versus varietal labeling."

***

In the first few years following the *Sideways* tsunami, the Santa Barbara region not only saw a spurt in the growth of Pinot Noir but it also saw the expansion and development of a new appellation. This appellation for the fog-bound, windswept western edge of the Santa Ynez Valley was established in 2001, but it's doubtful back then even Miles knew of its existence. The area is called the Santa Rita Hills, but when Viña Santa Rita, one of Chile's largest and oldest wineries (established in 1880), threw a legal fit and threatened international scandal, Richard Sanford flew to Santiago and brokered an agreement—over a glass of wine, he says—that saw the appellation abbreviated to "Sta. Rita Hills."

There were wineries in the area then, but not like today. You might say there was a forty-year cycle between the area's discovery by Richard and Michael Benedict in 1971 and the gold rush that began in the early 2000s.

The appellation came into being because everyone involved noticed how different these wines were, even compared to other local wines, thanks to multiple soil types and multiple exposures.

Then a Burgundian discovered it. Our old friend and flying buddy, Étienne de Montille, settled on Sta. Rita Hills in 2017 for his Racines Wines brand. After an extensive search of the West Coast, including Oregon's Willamette Valley, the Sonoma Coast, and the Santa Cruz Mountains, he and his partners (Brian Sieve, his American Chef de Cave at the domaine in Burgundy, and Champagne veteran Rodolphe Péters), determined this was "the appropriate terroir to make outstanding, fresh, mineral-driven, vibrant Pinot and Chardonnay. We want to make a California wine that expresses California terroir with a Burgundian vision."

What is the Burgundian vision?

"Respecting the soil without too much intervention, low alcohol, high acidity, minerally driven, fresh, vibrant, earthy. It is a stylistic and aesthetic vision. Sta. Rita Hills terroir is the key driver."

The land, "a beautiful hill with limestone" along the Highway 246 corridor, was acquired in 2017 and planted in 2019 and 2020. Already Étienne is noticing the difference between his approach and that of the locals. The Burgundian vigneron refuses to go for the bolder, more extracted style of some of his neighbors.

"We pick on the very early side," he declares. "We see some picking a month later than us from nearby vineyards. You will have a very different fruit profile and balance in the wine when you pick a month later."

California's Indian summers allow vintners, if they so choose, to pick riper fruit—which means a wine with saturated color, a lush mouthfeel, greater density, and alcohol in the 15 to 16 percent range. This plays into the ongoing debate in the American Pinot world about what is a true Pinot.

***

The movie caused a shift in how some vintners make the wine and in how some drinkers perceive its appeal. Let's talk to a few experts.

Rajat Parr did not grow up with wine. Born in Calcutta, India, he got his first introduction to fine wine from his uncle in London. Later his palate got fine-tuned and sophisticated with extensive travels around the world and in studies at the Culinary Institute of America in Hyde Park, New York.

Over time Raj became one of the world's best-known sommeliers, winding up as the wine director of The Mina Group in San Francisco and overseeing its eventual twenty-five outposts throughout the world. In October 2010 he published in collaboration with Jordan Mackay *Secrets of the Sommeliers*, which won a 2011 James Beard Foundation Award, Beverage category.

About the time *Sideways* was filming, Raj decided to get his hands dirty: he decided to become a farmer and winemaker. His initial foray into winemaking started in 2004 with the Parr label, sourcing fruit from prized vineyards in Santa Barbara County and producing wines at local wineries. In 2011, he launched Sandhi wines in partnership with an investor and the renowned winemaker Sashi Moorman, producing Sta. Rita Hills Pinot Noir and Chardonnay from purchased fruit.

Raj and Moorman established Domaine de la Côte in 2013, a forty-acre parcel that was initially developed in 2007 by Moorman in Sta. Rita Hills.

"We had no idea when we planted here how it would be," says Raj of this extreme western part of Sta. Rita Hills. We are standing on a hilltop overlooking a patchwork of six vineyards of Pinot Noir perched at an elevation of five hundred feet above the Santa Ynez River.

Raj picks up an ashen-colored silica rock, light as paper and porous. It's this soil that gives the wines elegance and finesse. The soil is noted for producing acid-retaining, mineral-rich wines.

A Burgundy aficionado who makes Pinots in California and Oregon, Raj practices minimal intervention—no additives such as enzymes, acids, or sugar, all-wild-yeast fermentation in cement tanks, and neutral oak barrel aging. His philosophy: "Let the grapes do their magic." His wines possess elegance with finesse. There's restraint combined with a riot of fresh flowers—violets, rose petals, geraniums layered with earthy mushroom and wild game notes.

The Sta. Rita Hills AVA more or less exploded around 2007, about the time Domaine de la Côte was planted. Today, the Sta. Rita Hills and Santa Maria Valley AVAs are prime Pinot country: the earthy, dark-fruit Pinots from Sta. Rita Hills and the cherry-loaded, spice-laced, perfumy Pinots from Santa Maria Valley.

Wes, who arrived in Santa Rita in 1994, today works in Santa Maria. He sees a difference in the two valleys:

"We make classic purist Pinot, especially with Santa Maria Valley's older vines, older clones. We make a more elegant, restrained, juicier, spicier wine. To make the new style of Pinot—bigger, richer, riper, darker, more intense, more extracted—that's kinda Sta. Rita Hills, a bolder and masculine style.

"Pinot purists still love what we do in Santa Maria Valley because of the elegance, restraint, loaminess, openness, and the perfume of these wines. To me, overt extraction in Pinot Noir robs the wine of its perfume. If you want to smell the place, vintage, vineyard, the fruitier and riper you make wine, the further you're moving away from that paradigm."

Dick Doré sees the comparison between the valleys in terms of weather patterns.

"Santa Maria Valley is warm winters cool summers," he says. "Sta. Rita Hills has cool winters and warm summers. They bud out and harvest at about the same time. Santa Maria Valley is red fruit—cherries, raspberries—and as Jim Clendenen always said, 'spice'—cinnamon, cloves, cardamom. Sta. Rita is dark fruit—blackberries, blueberries, earth, more Burgundian, more serious. Robert Parker called it 'underbrush.'"

This goes to the complexity of California Pinot Noir since *Sideways*. People used to quaffing Cabs and Syrahs, not to mention Petit Sirah or Tannats, find a purist Pinot "too light." So they go for the bigger, richer wines.

"One grower I work with has some magnificent Grenache," says Bryan. "We'll taste his Grenache and he'll look at me and say, 'That's the Pinot Noir that people want.' It's built around intense fruit and cranberry. Whereas our Pinots are more earthy, foresty."

This interesting comment goes to the heart of a debate among Pinot producers and lovers about the permissible boundaries of style. How full-bodied, rich, dark, and dense can a Pinot be before it becomes untrue to the variety? If you make a Pinot for a Grenache or Syrah lover, is that a true Pinot? Conversely, are the light-tinted, transparent Pinots with spice and intense floral notes like the authors encountered years ago in Oregon and Burgundy now a liability in the marketplace?

Let's throw a real monkey wrench into this debate and bring in climate change. Everyone has noticed how Burgundies are becoming more and more like New World Pinots.

"I think we can say Burgundy is no longer Burgundy and Oregon no longer Oregon—not the way we remember them in the eighties and nineties," says Wes. "We've already seen the terroir of specialty regions up around the 45th parallel where big temperature swings are massively influenced by climate change with hotter summers and more ripeness. These regions are producing wines with California sunny ripeness. Alcohol ticks up with more ripeness and fruit, more of a New World style. Master somms are having a hard time recognizing the difference between Burgundy and California now."

Raj agrees: "Burgundy is not cool anymore. In Burgundy, it's hard to find wine under 14 percent [alcohol] now. So the world is changing. You're finding lower alcohol wines in California than in Burgundy. It makes farming very tricky."

Some folks are picking their Pinot earlier to offset climate change.

"The trend in Burgundy, Germany, and certainly with people we know is toward picking earlier to deal with global warming and make the kind of wine Pinot Noir was originally meant for, which is a much more transparent style," said Oregon's David Adelsheim.

What needs to happen, he insists, is for the world of Pinot—meaning producers globally—to seriously research how this grape will survive a warming trend in

Miles's "tucked-away corners of the world." If certain clones can ripen earlier, then there must be clones that ripen later. Are there rootstocks that can ripen two weeks later than standard rootstocks?

"The research around the world hasn't happened but needs to happen for places like Oregon and California," he says. "There are things we can do without abandoning Pinot Noir."

***

Before we dig much deeper into what has and has not happened to Pinot and Merlot since the movie, let's introduce another vigneron, a global winemaker and Burgundian who is building an empire in California, one founded originally on Pinot.

Jean-Charles Boisset was born in the lovely village of Vougeot. As a boy, he woke up to a view of Clos de Vougeot. "So my life has been Pinot Noir," he chuckles.

He may well be the most colorful vintner in America, with an emphasis on *color*. Take his signature red socks, which fit snugly inside Christian Louboutin shoes, or his bold designer jackets adorned by gold brooches, part of the jewelry collection he designs. Then there's the high-octane energy so powered by his unbridled enthusiasm that you get high without a sip of his wines.

But don't let his charismatic and flamboyant image fool you. The astute vigneron smartly oversees a global wine empire. Jean-Charles Boisset, or JCB as he is fondly known, is president of Boisset Family Estates, with a portfolio of wineries located in the most prestigious terroirs, from Burgundy, Rhône Valley, and the South of France to California's Napa and Sonoma Valleys. The Bordeaux-style house, Raymond Vineyards, JCB's ninety-acre flagship estate in Napa Valley, was acquired in 2009.

"Three things happened to Pinot Noir because of *Sideways*," says JCB. "People discovered they liked that taste and didn't really know what to do with it. They discovered they liked that flavor profile. And they discovered the sex appeal of Pinot.

"The third is the important one—that it's a sexy, great wine you can never really expect the same every time. In great [wine] places like the Rhône Valley, Châteauneuf-du-Pape is very consistent with animalistic, leather, caramel feel, right? Pinot, though, is like a gorgeous woman who dresses up beautifully and always surprises you with her feel and scent. You know you're going to love it, but you see her differently each time."

Thus, Pinot Noir is positioned very well by the movie as a wine for seduction, JCB insists. Movies in the old days or TV shows such as *Sex and the City* equated spirits—cocktails, typically hard alcohol—with sex. *Sideways* introduced wine as seductive.

The flamboyant vintner Jean-Charles Boisset, proprietor of Boisset Collection, Napa Valley, California AUTHOR'S COLLECTION

Winemaker Adam LaZarre, Paso Robles, California
AUTHOR'S COLLECTION

Vintner Rajat Parr at his Bloom's Field vineyard in Sta. Rita Hills, Santa Barbara County, California AUTHOR'S COLLECTION

Bryan Babcock, owner, Babcock Winery & Vineyards, Sta. Rita Hills, Santa Barbara County, California AUTHOR'S COLLECTION

"I was raised in France on American movies—especially Westerns. There was always a glass of whiskey that identified American film and television with hardcore spirits. Finally, now is the first movie I've seen to talk about wine for ninety minutes. There are no alternative drinks."

***

Americans grew up drinking Coca-Cola. They then discovered spirits, and many still seek the big and flavorful in their adult beverages.

"When people drink richer, bolder wines, they come from drinking spirits in US," notes Raj. "People who started to drink Pinot Noir probably came from Cabernet first, right? So bigger Pinots satisfy people who drink Cabernet. Which is crazy. There's no similarity between those varieties. If you make a Pinot for a bigger drinker, you're just making a Pinot to fit a clientele."

He continues: "There's no way to make a *light* Cabernet. It doesn't happen. But Pinot you can make light or massive. That's just winemaking."

"With bourbon, people like the mouthfeel," notes JCB. "Brandies can be round, gentle, enjoyable if not too high in pH. The pleasure is in the mouth; it's not just an intellectual journey. The balance is better. Why red blends are so successful is balance. Balance is key. It's hard to achieve with Pinot. You can't blend it.

"Why Cab or Cab Franc is so successful is the attack is there—bang! American consumers loved those wines' powerful presence. With Pinot, they can be disappointed."

"There's a lot of good Pinot, but there's not that many *great* Pinots," says Raj of the Golden State. "A lot of Pinot tastes delicious, but is it special? That's very little. There are maybe twenty great sites in the whole state, if there are that many."

"I'm often disappointed with Pinot where the attack is so-so, the length average, then midpalate this and that," says JCB. "I would say six times out of ten I'm happy with a Pinot. With a Cab or red blend, nine times out of twelve I'm satisfied."

An often-overlooked aspect of the *Sideways* effect is that the vintages surrounding the film's production and release, 2003 and 2004, were two of the warmest years in California winegrowing. Those years were atypical for Pinot Noir because of the weather, which led to bigger, richer, higher-in-alcohol wines. So people drinking Pinot Noir for the first time thought this was what Pinot Noir was like.

Adam Lee, whose Siduri label (since sold) produces superb Pinots, and only Pinots, from many of the iconic, cool-climate regions along the West Coast, says this about the coincidence of these anomalous vintages with the release of the movie: "I think it led many of us, myself included, to make bigger, richer, more extracted Pinot

Noirs, even in years which weren't necessarily designed for that. It led all of us to a slight period of excess. Had *Sideways* come out two years earlier or two years later, what people would have come to expect would have been very different."

This then resulted in a winery association called IPOB (In Pursuit of Balance) that moved to a leaner style of Pinot Noir. "I think sometimes those wines were made with a purposeful stylistic bend in mind," remarks Adam. "Sometimes it worked in certain locations, but then certain years didn't lend itself to that style either. It was this pendulum that went a little too far in California into the ripeness style, and then we probably went a little too far in the leaner style."

As for where we are today, he smiles and says "We're in a better spot now than then."

***

So where are we now with Pinot Noir? No one better to ask than Sonoma winemaker Greg La Follette, who has made wine on five continents and even lectured about Pinot in Burgundy—"very cheeky for an American to do that, but they invited me."

"There's been a lot of advances in making Pinot Noir, but the most important advances are going to come in the vineyard," insists Greg. "The more winemakers get savvy to what's happening in the vineyard, the better Pinot Noir they're going to make."

So many winemakers, even makers of Pinot, either through egotism or poor training, insist they make "their wine" in the winery. Yet Greg maintains that the most important moments in Pinot Noir winemaking occur before a single berry makes it across the winery threshold.

A self-described cellar "hose dragger" in bib overalls with the weathered face of a man who has spent nearly seventy harvests in vineyards, Greg takes a break from his barrel room to sit with his young and thoughtful associate winemaker, Evan Damiano, at Marchelle Wines in Sebastopol, Sonoma. He takes us through his methods of coaxing the best Pinot Noir he can from vineyards he's worked with for years.

When those berries do cross the threshold and the juice gets in barrels, Evan says their process then is "to treat each barrel as a child. Listen to it. See what it needs. Does it need more time? Does it need more support? If it starts going off the rails, make adjustments."

"Pinot is a wine-growers' grape, meaning if you get it ripe in the vineyard, all you have to do in the cellar is nurture it," agrees Matt Revelette, winemaker at Siduri since Adam sold it. "You're not going to improve things a lot if you planted in the wrong spot."

More and more producers are taking better care of their soils and environmental footprints than twenty years ago. Biodiversity, land preservation, and regenerative farming may be the new buzzwords, but many winemakers and proprietors have made them marching orders. And the wines are better for it.

One morning we walk through La Crema Estate's Saralee Pinot Noir vineyard in Sonoma with winemaker Eric Johannsen and sustainability analyst Alexandra Everson. Within an experimental five acres, there are a mobile chicken coop (on a tilt so the eggs can roll down) and a flock of sheep rotating through the various rows reducing the cover crops. Several trials are going on simultaneously.

"We want to be good stewards of our land, reduce our carbon footprint, and have healthier soils," Alexandra tells us. "These vineyards are getting older, and we're learning from trials with animals to come up with better farming practices."

Regenerative farming improves soil health by moving some of the excess carbon from the atmosphere (where it's not needed) into the soil (where it is needed). The carbon within the soil is rich in nutrients, can hold water better, and creates an environment for favorable microorganisms to populate.

La Crema creates compost on-site from the pomace, which is the grape seeds and skins left over from the winemaking process, and applies it back in the vineyard. The estate has installed a rainwater capture system that in a year can collect eighty thousand gallons of rain.

Regenerative farming has joined organic and biodynamic practices to bring some sanity back to farming. Once dismissed as gimmicks, many farmers have come to realize that far from being new things, they are very old things—the way farming was practiced for many millennia until we moderns introduced chemistry in the form of pesticides, fertilizers, and other tactics, the very things that nearly killed the great vineyards of Burgundy. It can admittedly be complicated and expensive to worry about such things.

"Pinot is very prone to disease, so you have to be careful what you're spraying and how you take care of the vineyard," says Raj. "That's why organic farming gets tricky on the coast. In cool and wet climates, it's just a lot more work. With conventional farming, you spray a systemic spray that goes into the plant and makes the plant inside and out resistant to disease. But if you're organic, you have to spray every week or every other week depending on how much mildew pressure you have. If you have a massive vineyard and it's conventionally farmed, it's a lot less work, but you can taste the difference in the wine."

***

The original vine plantings in California and Oregon came from a desire by planters to emulate the greats in Europe, Clare believes. Pinot planters looked fondly to Burgundy; Cab and Merlot plantings were done by Bordeaux aficionados.

"In California, that period of emulation is over," she insists. "People aren't planting Pinot Noir to match what's going on in Burgundy. They have found [their] own character [in the New World] and are proud of it, hence the proliferation of Pinot Noir in appellations up and down the coast. Americans were never enamored of Burgundy—they never even tasted it. For a long time, less than 10 percent of the wine available in the United States came from elsewhere.

"Consumers are looking for a winemaker they like, a grape they can pronounce, a style that's not going to attack them and be aromatic, fruity, delightful. The American population loves a brand it can be loyal to."

Pinot Noir has always been a top-tier wine in Europe. *Sideways*, however, helped bring the wine to the world stage.

"The popularity is now global," agrees Raj. "Everyone is looking for places to plant Pinot—New Zealand, Australia, California, Patagonia, South Africa. In this global movement, there are only small plots that are special, though."

There's the rub—"only small plots are special." So Pinot Noir, now firmly established globally, may continue to be elusive and inconsistent. It may continue to tantalize and frustrate as it gets planted in good soils, but not special ones.

Pinot became "a little too commoditized for a while" after *Sideways*, says Matt. The usual speculators were looking to exploit something suddenly popular and producing bland wines. He also thinks growers have become more disciplined in site selection. Farming too has zeroed in on rootstock selection, especially looking for drought tolerance.

As far as Bryan is concerned the single biggest issue in winemaking today is oversupply. "There's so much wine worldwide for God's sake. That's why I got out of the rat race," he says. He reduced his annual production from twenty-five thousand cases to five thousand cases.

"I buy a lot of bulk wines," says winemaker Adam LaZarre. "I'm starting to see a lot of Pinot in abundance. A few years ago, half a million gallons of Santa Barbara Pinot Noir was on the bulk market. No smoke problem, they just had a huge bumper crop. Pinot Noir prices are down, not to the pre-*Sideways* level but pretty close."

"I don't know if that [oversupply] diluted the quality of Pinot Noir overall though," says Bryan. "The market itself is always improving the quality of the wine, because people making bad wine go out of business. I would say the quality of wine at a lower price point is better than it was twenty years ago. Pinot Noir is now great."

JCB would like to believe this but is unsure. He comments, "I feel after *Sideways* Pinot went crazy, then a little flat, and now it's challenging. It's doing okay. But we're not doing great. I feel we need another movie. The *Sideways* effect was great, but we need another *Sideways*—maybe in Sonoma this time."

Pinot Noir from New Zealand

Wine grapes were brought to New Zealand by British missionaries from Australia in the 1800s. It wasn't until the 1970s that the islands would see a rapid boom, especially in Marlborough's hallmark Sauvignon Blanc, which firmly secured New Zealand wine's prominence internationally.

From some nine hundred acres in 1960 to a current ninety-eight thousand acres, the country has seen a dramatic upswing and found a fan base for Pinot Noir.

The two strongholds for Pinot Noir are Marlborough, which accounts for some 60 percent of vineyards planted in New Zealand, and Central Otago, an inland wine region (most other regions are coastal) that is home to the southernmost vineyards in the world, with Antarctica only 2,500 miles due south. Being in a nation of just over five million people, the New Zealand wine industry depends upon exports. Some wineries, though, may be too small to have US distribution, but many can ship directly to consumers.

Pinot Noir from Marlborough, Central Otago, and North Canterbury, New Zealand © ZW IMAGES

Our suggested Pinots of New Zealand:

Burn Cottage (Central Otago)
Dog Point Vineyard (Marlborough)
Greywacke (Marlborough)
Pyramid Valley Angel Flower (Waikari North Canterbury)
Quartz Reef Bendigo (Central Otago)

Rex Pickett's New Zealand suggestions:

The Boneline Wai-iti (Waipara Valley)
Pisa Range (Central Otago)
Prophet's Rock (Central Otago)
Three Miners Vineyard (Central Otago)
Two Paddocks (Central Otago), owned by actor Sam Neill

20

The Alchemy of *Sideways*

"Twenty years later, many films I've done don't continue having relevance in our culture," says Sandra Oh. "*Sideways* still holds up in a lot of ways in beginning to investigate what that straight white male friendship is. The movie was a real window into that."

"Alexander is an artist," says David Lonner. "He made, in my mind, one of the great romantic, serious, intelligent, joyful, painful romantic comedies of the century. It was the best experience I've ever had in the movie business. You can have the greatest script, the greatest director, the most incredible cast, and the movie sucks. You realize there is an alchemy sometimes. It happened on this movie."

In medieval times alchemy concerned the supposed transformation of matter, a protoscientific tradition that saw many instruments and laboratory processes handed down to modern science, yet an air of mystery and the occult clung to the practice.

Like the alchemists of old, moviemakers have studied and applied the art form and sometimes do achieve the kind of alchemy David speaks of, the kind that lifts a popular film out of its time and place to have an ongoing relevance to our culture, to become a "classic." How this happens is a mystery. There's no science here, no formula or recipe that if followed correctly ensures that result.

Alexander has made films every bit as good as *Sideways*, and his humanistic, bitingly funny, and hauntingly melancholy comedies will stand the test of time. Yet *Sideways* is *Sideways*.

What he understands in all his films is that while dubious intentions, physical risk, and explosive comedy make for fine stories, the thing that makes a story last is character. These particular characters in *Sideways* belong in the pantheon of American film comedy.

Comedies portraying male misbehavior and alcoholic indulgence are not infrequent in American cinema. But as one would expect in an Alexander Payne film, he looks at these unruly male characters as wonders of misplaced machismo, self-pity, and human frailty. Pathos mixes with the pratfalls, and comedy is found in the deepest despair. These are straight white males at the turn of a new century, having learned little from the previous one, and finding in friendship a false sense of security to act in ways they might not dream of doing alone. Wine appreciation masks alcoholism, just as womanizing masks a failure to mature.

Miles is a failure in his own mind. Jack is a success at what a more mature individual would shun. The buddies in this buddy movie are a mess, and friendship helps them only so far. Miles's advice to Jack is astute, but Jack never really listens. Jack's counsel to Miles is sincere and smart—"No going to the dark side"—but Miles takes it as disapproval to fuel his grand-scale pity party.

"Miles is not a courageous character at all," notes Sandra. "The fact he has a little shift in his concept of self-worth is what the film ends up being about."

Miles gets, albeit briefly thanks to his pal's flagrant conduct, a sensual, intelligent woman looking for love, too. He encounters, albeit with remorse, his ex without fear or trembling. He sees the wisdom in Maya's enjoyment of wine rather than abuse of it. And he returns to the scene of many crimes with hope in his heart and perhaps, as Virginia suggests, for a cup of tea and sympathy.

Looking back across two decades of American cinema, which saw the rise of superhero movies—which no less than Martin Scorsese has said are closer to theme parks than they are to movies—and the industry's metastasizing aversion to independent film and movie theaters in general, audiences are left to wonder if a *Sideways* is still possible.

Alexander's *The Holdovers*, produced by Gran Via and the Harvey Weinstein-free Miramax, was released by Focus Features on October 27, 2023. Reviewing a film seeking melancholy humor and human triumph over life's tribulations but *not* a franchise or a tentpole, *IndieWire*'s David Ehrlich posited that Alexander "takes great pleasure in defying every impulse of modern cinema."

This is somewhat backward, as the filmmaker hasn't changed his approach to classic American cinema one iota since he made *Citizen Ruth* in 1996. Apprehensive studio executives and changing technology have transformed the American movie machine, almost unrecognizable in its embrace of superheroes and endless franchise movies.

As he has throughout his career, Alexander, who commissioned the screenplay by David Hemingson based on an obscure French picture, *Merlusse* by Marcel Pagnol, focuses on damaged humans who are struggling with life's vicissitudes. The story takes place at a New England all-boys boarding school over the Christmas holidays in 1970, where the campus pretty much empties except for the "holdovers," kids with nowhere to go for one reason or another. This gets further reduced to three people when one wealthy family whisks the students away for a ski trip with parental consent for all but one student.

This trio then, a cranky history teacher (Paul Giamatti), seriously disliked by students and faculty alike; a heartbroken head cook (Da'Vine Joy Randolph), who has lost her only son in Vietnam; and a troublemaking albeit supersmart youth (Dominic Sessa in his first film) pretty much abandoned by anyone he might call family, will figure out how to create a semblance of a new family over these several days. But it's in the revelations that slowly seep out from the interchanges among the threesome and the reversals of fortune they encounter that Alexander finds his comic drama. He smoothly skates over the potential pitfalls of sentimentality and simplistic answers to get beyond first impressions and, as he did in *Sideways*, get an audience to fully appreciate the what, where, and why behind the plights of his characters. The movie begins with the well-worn types in any campus movie, yet these evolve into fully realized human beings stressed by circumstances as beyond their control as any hero in Greek tragedy.

The reunion of Alexander and Paul, contemplated by both for many years, enriches this sagacious comedy that even hints at a spiritual connection between the distressed Miles of twenty years earlier and this cantankerous professor. Last seen stuck in a miserable teaching post, Miles might well have eased up on his drinking—the professor does get tipsy on occasion during the holidays—only to find himself stuck in a career in education rather than literature. The bitterness that oozes into his trenchant commentaries and clashes with students and the headmaster do sound like Miles's critiques of wines he finds less than transcendent, and his fatalism is every bit as rigorous as the professor's.

***

It's heartening to know that Alexander is not all lost in the daunting landscape of streaming, limited series, and multiple platforms. Alexander is open to working with streamers. He nearly did a film for Netflix a few years back. The untitled film, which was to star Mads Mikkelsen, was very loosely based on a *New York Times* article by the Norwegian author Karl Ove Knausgaard.

"Netflix promised me a two-to-three-week theatrical window. So they were open to doing films that studios are not going to finance," he says. In this particular instance, the plug was pulled four days before shooting was to commence.

Another lost project occurred more recently, when he was lured into helping create a limited series for himself based on a novel. Another writer wrote two of what would be eight episodes.

"It was good and very funny. We were ready to pitch it to key buyers—Amazon, HBO, Netflix, FX, Hulu. We met and it was all hugs and kisses. But only one buyer bit. We had a Zoom pitch meeting—the first time in my life I've ever pitched. They said, 'It was wonderful, fantastic, and we'll get right back to you.' We never heard from them.

"I asked my agent and producers in TV, 'Is that how things go in TV—you spend a bunch of time and money developing something and getting it out to buyers and if nobody goes for it that's it? It's dead?' 'Pretty much,' they said.

"I'm sticking with features. This TV stuff is for the birds."

Among other ideas, Alexander is toying with tackling one of the American cinema's oldest and most neglected genres, the Western, which he is working on with *The Outsiders* screenwriter David Hemingson. He and Jim are also collaborating on an unusual project that, if it were made, would be in French.

A few years ago, a producer approached the duo about a *Vanity Fair* article about rival antique chair dealers in Paris.

"It has boffo box office written all over it," Jim jokes.

Alexander has enough projects backed up that a preferred method of getting a writing project on its feet is to farm out the project to other writers and then rewrite based on their drafts. For this project, the two settled on Thomas Bidegain (Jacques Audiard's brilliant *A Prophet*, *Rust and Bones*) and Noé Debré (Tom McCarthy's *Stillwater*).

"They wrote a draft or two, which are fine, but to get our voice into it rather than give notes for another draft I thought let's have Jim and me take a pass at it," notes Alexander. "It's going to be in the French language, too, so we'll give back our draft to the guys to sprinkle French perfume all over it, to Frenchify and translate it."

Alexander is almost allergic to sequels, but then again . . . *Election* novelist Tom Perrotta wrote a sequel called *Tracy Flick Can't Win*.

"So a deal is in place for Jim and me to write and me to direct and Reese to star in a sequel, but I just haven't gotten to it," he says. "The new novel, as fine as it is, will need a lot of invention to make a motion picture—at least one that I would make."

He is completely comfortable in this brave new world of technology because he sees his job as technology-proof. "No matter what the technology or distribution, it will always come down to what's the story, who are the actors, where do the actors stand, where do we put the camera, how do we cut it together?—that's never going to change. From 1903 till now that stuff never changes."

At the 2023 Venice Film Festival, where indie filmmaker Richard Linklater premiered his film *Hit Man*, he gave an interview to *The Hollywood Reporter* where he wondered if the independent film movement isn't "gone with the wind—or gone with the algorithm. Sometimes I'll talk to some of my contemporaries who I came up with during the 1990s, and we'll go, 'Oh my God, we could never get that done today.' So on the one hand, selfishly, you think, 'I guess I was born at the right time. I was able to participate in what always feels like the last good era for filmmaking.' And then you hope for a better day. But, man, the way distribution has fallen off. Sadly, it's mostly just the audience. Is there a new generation that really values cinema anymore? That's the dark thought."

Bobby Roth may be the ultimate indie film guy. While he pays his bills in commercial television, he has made independent films across seven decades. So, we ask him, is there even a thing called independent film anymore?

"I don't think so," he says frankly. "If there is, it's independent of me.

"You can't make little movies. I don't know what the movement is unless it's very much one-offs. [Indie filmmakers] make one movie and that's it. Filmmakers just want to do series or streaming."

Netflix's policy of buying global-only rights has had repercussions on smaller indie films, including the loss of a theatrical release. That led to a decline of indie films in movie theaters, which was followed by the pandemic shutting many theaters down. In recent years all streaming services have cut back on acquiring indie titles, making it difficult for indie fare to find homes. Netflix has shown signs of changing its policy by buying some films for less than the entire world, but it remains to be seen how this will play out.

***

After our interviews with Alexander Payne concluded, in an email exchange between the authors and the filmmaker we wondered if *Sideways* was even possible today. A few evenings later, Alexander sent this email:

> I'm intrigued by your question as to whether *Sideways* would be "even possible today."
>
> I'd like to think that it would, if such a marvelous piece of source material plopped again into my lap. Rex's novel, like Kaui Hemmings's novel later for *The Descendants*, was a living, breathing chronicle—largely autobiographical. It had

been lived. Those are the stories that appeal to me. I've spent my career still trying to make seventies movies.

Don't forget keeping the budget low. Very important. *Sideways* cost only some $15.8 million, *The Descendants* and my new one *The Holdovers* both in the low-to-midtwenties. Hell, *Nebraska*—ten years after *Sideways*—was only $13 million or so. This is where freedom lies. And, sadly, perhaps where humanism is still able to lie.

I'm reminded of an anecdote of John Huston meeting Luís Buñuel. Huston asked don Luís how it was that he could fashion such eccentric masterpieces as *Los olvidados* and *Viridiana*. Don Luís said, "How much are you paid, and how much do you think I am paid?"

Too many filmmakers say with their films, "Look at me." I like the filmmakers who say, "Look at this story. Look at these people."

I will spend the rest of my career trying to master classical American filmmaking.

Alexander

Acknowledgments

For this book, the authors spoke to filmmakers and winemakers alike, and in a few instances relied upon long-ago interviews. So, in a sense, we've been unknowingly working on this for much of our professional lives.

At the head of the parade of those to acknowledge and thank is Alexander Payne, who generously devoted many hours talking with us on Zoom, not just answering our myriad questions but explaining his filmmaking philosophy and ideas about comedy and, specifically, classical American filmmaking. His enthusiasm for our project led so many others—his actors, key crew members, and colleagues—to talk at length in interviews.

The film's producer, Michael London, not only gave us an interview while suffering a raging head cold but facilitated contact with other significant people involved with *Sideways*.

Rex Pickett, who wrote the novel on which the movie is based, not only granted us a lengthy in-person interview in Napa Valley but tolerated numerous follow-up emails and Zoom chats over many months, to say nothing of his fine personal photographs.

The film people we spoke to, in alphabetical order, are Thomas Haden Church, Paul Giamatti, Steve Gilula, Rolfe Kent, David Lonner, Virginia Madsen, Sandra Oh, Phedon Papamichael, Peter Rice, Bobby Roth, Jane Ann Stewart, Jim Taylor, Mark Tchelistcheff, and Kevin Tent. Alexander's then assistants Evan Endicott and Rachel Fleischer proved invaluable for their candid photography and insights into the shoot, as did his former intern Brian Beery.

Many of the winemakers we spoke to are old friends, but we were delighted to meet many new ones not only over Zoom and in phone interviews but in person in the Santa Barbara area, Oregon's Willamette Valley, the Sonoma and Napa Valleys, Santa Lucia Highlands in Monterey, and Anderson Valley in Mendocino.

These include David Adelsheim, Joe Alarid, Kyle Altomare, Renée Ary, Rob Astrin, Bryan Babcock, Jean-Charles Boisset, Morét Brealynn, Tresider Burns, Melissa Burr,

Deborah Cahn, Chris Carpenter, Maggie D'Ambrosia, Evan Damiano, Étienne de Montille, Aubert de Villaine, Richard Doré, Gary Eberle, Michael Etzel, Alexandra Everson, Marc Goldberg, Wesley Hagen, Gina Hennen, Chris Hyde, Larry and Beta Hyde, Eric Johannsen, Kathy Joseph, Michel Lafarge, Greg La Follette, Adam R. LaZarre, Adam Lee, Jason Lett, Bob Lindquist, Sam Marmorstein, David Millman, Jessica Mozeico, Tom Myers, Kimberlee Nicholls, Frank Ostini, Rajat Parr, Harry Peterson-Nedry, Mark Pisoni, Matt Revelette, Zac Robinson, Richard Sanford, Katie Santora, Susan Sokol Blosser, Hailey Trefethen, and Doug Tunnell. Add to this long-ago interviews with Jim Clendenen.

A special thanks to the late Becky Wasserman, the renowned Burgundy importer headquartered in Beaune, for arranging many introductions on our visits to Burgundy.

Dr. Stephen Cuellar, professor of economics and wine business at Sonoma State University, and Master of Wine Clare Tooley, VP of guest experience at the Boisset Collection, offered their professional expertise.

Our literary agent, Lee Sobel, helped turn the idea for this book into a reality. It's been a pleasure to work with our editor, John Cerullo, and our photo editor, Barbara Claire, at Rowman & Littlefield.

Many people in the wine business and marketing/communications helped us immeasurably in arranging winery visits, opening doors to winemakers and owners, and securing bottles for photography. Among these are Ana Cahaus, Barbie Chiu, Megan Conway, Sam Dependahl, Amy Freeman, Sonia Ginsburg, Anna Hanson, Erin Inman, Miriam Jonas, Ranit Librach, Gwen McGill, Liddy Parlato, Emily Peterson, Kim Stemler, David Strada, and Jamie Tobin.

We also salute those companies and wineries who have greatly aided our research: andcopr, Charles Communications Associates, Clendenen Lindquist Vintners, Colangelo PR, Double Forte PR, Foley Family, Jackson Family Wines, J.A.M. PR, O'Donnell-Lane, and Vieux Communications.

Our Paso Robles–based photographer was Dianne Zwick, greatly assisted by Glenn Williams. A special thanks to the Allegretto Vineyard Resort for lending their enchanting landscape for our photography. And a special thanks to supportive friends Barbara and Tom Corlett.

Notes on Sources

Books

Boidron, Robert. *Le Livre du Pinot Noir*. Paris: Lavoisier, 2016.

Bonné, Jon. *The New California Wine: A Guide to the Producers and Wines behind a Revolution in Taste*. Berkeley, CA: Ten Speed Press, 2013.

Broadbent, Michael. *The Great Vintage Wine Book*. New York: Alfred A. Knopf, 1982.

Farber, Manny. *Movies by Manny Farber*. New York: Hillstone, a division of Stonehill, Publishing, 1972, a reprint of his *Negative Space*, originally published by Praeger Publishers in 1971.

Graham Jr., Otis L., et al. *Aged in Oak: The Story of the Santa Barbara County Wine Industry*. South Coast Historical Series Graduate Program in Public Historical Series. Santa Barbara: Santa Barbara Vintner's Association, University of California–Santa Barbara, 1998.

Haeger, John Winthrop. *North American Pinot Noir*. Berkeley: University of California Press, 2004.

Haeger, John Winthrop. *Pacific Pinot Noir*. Berkeley: University of California Press, 2008.

Kerr, Walter. *Tragedy and Comedy*. New York: Simon and Schuster, 1967.

MacNeil, Karen. *The Wine Bible*, 3rd ed. New York: Workman Publishing, 2022.

Potter, Maximillian. *Shadows in the Vineyard*. New York: Twelve, Grand Central Publishing, Hachette Book Group, 2015.

Roberts, Jerry. *The Complete History of American Film Criticism*. Santa Monica, CA: Santa Monica Press LLC, 2010.

Smith, Lory. *Party in a Box: The Story of the Sundance Film Festival*. Salt Lake City: Gibbs Smith Publisher, 1999.

Vachon, Christine. *A Killer Life: How an Independent Producer Survives Deals and Disasters in Hollywood and Beyond*. New York: Simon & Schuster, 2006.

Vidor, King. *A Tree Is a Tree*. New York: Longmans, Green and Co., 1954.

***

Articles and Papers

Biederman, Patricia Ward. "A Reel Hit: Filmmaking: After a Screening of His Student Movie, 'The Passion of Martin, Alexander Payne Was the Talk of the Town." *The Los Angeles Times*, July 8, 1990.

Brezeski, Patrick. "Richard Linklater on His Comedy Thriller, 'Hit Man' and Why Indie Movies Might Be 'Gone With the Algorithm.'" *HollywoodReporter.com*, September 5, 2023.

Cuellar, Steven S. "The Sideways Effect: A Test for Changes in the Demand for Merlot and Pinot Noir Wines." *Journal of Wine Economics* 4, no. 2 (January 2008), 219–32.

Hartke, Kristen. "'The Sideways Effect': How a Wine Obsessed Film Reshaped the Industry." *The Salt*. NPR.org, July 5, 2017.

Honeycutt, Kirk. The author while the film critic for the *Daily News of Los Angeles* reviewed and wrote extensively on regional and independent films in the 1970s and 1980s. These articles form much of the background to the section about the American independent film movement. While this publication still exists (barely), it has no digital archive going that far back. He also pursued this subject in articles written for *The New York Times*, whose archive is searchable: "Mavericks of the Movies—Regional Filmmakers," November 9, 1980; "Women Film Directors: Will They, Too, Be Allowed to Bomb?," August 6, 1978.

Kopman, Lewis. "WWC21—Charles Coury Vineyard, Oregon." JancisRobinson.com, August 14, 2021.

Noble, Greg. "Death, Wrath and the Real Story behind Rose's Ban." WCPO.com, December 16, 2015.

Index

Page references for photos are italicized.